D0890242

SECURITY RISK ASSESSMENT AND MANAGEMENT

SECURITY RISK ASSESSMENT AND MANAGEMENT:

A Professional Practice Guide for Protecting Buildings and Infrastructures

By

Betty E. Biringer
Rudolph V. Matalucci
Sharon L. O'Connor

John Wiley & Sons, Inc.

Library of Congress Cataloging-in-Publication Data:

Biringer, Betty E., 1952-
    Security risk assessment and management: a professional practice
guide for protecting buildings and infrastructures / by Betty E.
Biringer, Rudolph V. Matalucci, Sharon L. O'Connor.
        p. cm.
    Includes bibliographical references and index.
    ISBN-13: 978-0-471-79352-6 (cloth)
    ISBN-10: 0-471-79352-3 (cloth)
    1. Buildings--Security measures. 2. Risk assessment.
I. Matalucci, Rudolph V.   II. O'Connor, Sharon L.   III. Title.
TH9705.B75 2007
658.4'7--dc22
                                                    2006023980
Printed in the United States of America.

10 9 8 7 6 5 4 3 2 1

... to the volunteer men and women of the Interagency Forum for Infrastructure Protection (IFIP) who gave of their vision, commitment, and determination to protect and secure our critical national infrastructure, long before the events of September 11, 2001.

Contents

Figures

Tables

Preface

Our purpose is to provide a professional best practice guidebook for engineers, architects, security specialists, law enforcement and emergency management officials, and managers who are responsible for secure and safe workplace environments for occupants and owners of buildings and supporting infrastructures. To protect against malevolent acts against buildings and their occupants, a security risk assessment and management process must be useful for: (1) identifying a regional and site-specific likely and credible *threat spectrum*, and subsequent development of a design basis threat, (2) evaluating *consequences* including loss of life and property, economic impact, and loss of any symbolic value and public confidence, and (3) assessing the *ineffectiveness of the physical security and cyber-security systems* against the threat and identifying any site-specific vulnerabilities in the security system.

Our intent is to provide a systematic and robust security risk management approach about "how to" perform a complete risk assessment that assists the project owner and manager in deciding "why to" either accept the calculated risk or reduce the risk to a more acceptable level. The procedure for a viable risk reduction strategy is then addressed through the application of performance-based alternative security upgrades or consequence mitigation measures.

The expanding national needs for adequate protection of the public and real property against malevolent acts of terrorism, the availability of a variety of alternative security measures, and the demonstrated performance of these protective measures has created a paradigm shift in how currently prescriptive-based building codes might be limited where they are applied to the construction industry. Clear definition of basic security standards, especially for nongovernmental and commercial facilities, are currently minimal or nonexisting under malevolent threat conditions, and protection is usually dependent on a hypothetical and suspected threat and postulated protective system vulnerabilities.

The need for a validated means to determine these new security requirements becomes more apparent following the events of September 11, 2001. A rigorous application of a security risk management approach seemed appropriate and advisable in the national interest. This security risk management approach has been applied to some federal facilities and can effectively be used to justify any requirement for a level of protection. This approach can also be used to demonstrate the performance of security upgrades or consequence mitigation measures and to ensure cost-effective return on capital investments. A viable risk management process, where used appropriately, also acts against any form of false security that might result from a poorly conceived and inadequately justified and engineered upgrade and mitigation plan.

This guide book adapts the robust security tools and techniques developed by the Department of Energy's lead national security laboratories for use across our homeland. We have been motivated, therefore, to apply our lessons learned for the application of a systematic process followed in this text that shifts from the widely applied compliance-based security upgrade procedure to a futuristic applied performance-based system evaluation using tested risk methodologies. No longer do we consider that applicable standard security codes alone are sufficient for use in implementation projects. Our hope for the future is that the use of a risk-management-based approach will dominate security evaluations and analyses required before any perceived or recommended corrective action is undertaken. This "best practice" guide book will hopefully provide the necessary guidance to professionals and further assist with the management of security risks through a step-by-step procedure that achieves risk reduction and adequate security performance.

Forms and additional resources are available online at www.wiley.com/go/securityrisk.

Acknowledgments

We wish to thank the technical reviewers, especially Ivan Waddoups and Greg Wyss, for their numerous helpful comments and suggestions that immeasurably improved the content and description of the process. We are grateful to Elizabeth Affeldt and P. Rebecca Baca for their prompt and careful editing of the text and most especially to Jackie Ripple of Tech Reps, a division of Ktech Corporation, for her diligence and timeliness in preparing the text and figures for publication. We appreciate Tommy Woodall and Carla Ulibarri at Sandia National Laboratories for their management support of this work.

We are grateful for the visionary guidance and support given by all the early contributors to the risk assessment methodology process development. The prototype infrastructure security risk assessment methodology was developed by Sandia National Laboratories for the Interagency Forum for Infrastructure Protection (IFIP), a consortium of federal agencies. Included as charter members of the IFIP are the US Army Corps of Engineers, the Bureau of Reclamation, the Tennessee Valley Authority, the Federal Bureau of Investigation, the Bonneville Power Administration, the Western Area Power Administration, and Sandia National Laboratories. Without the financial and intellectual contributions of these member agencies, the rigorous, replicable analytical process described in this book would not exist today.

William K. Paulus, formerly of Sandia National Laboratories, must be recognized for his technical contributions throughout the risk assessment methodology process development. His contributions to the application of fault trees, risk evaluations and calculations, consequence table development, and threat assessment procedures are invaluable.

We gratefully acknowledge Dennis Miyoshi, director of the Security Systems and Technology Center, Sandia National Laboratories, for his persistent support, vision, and financial support, which made the preparation of this manuscript both technically rewarding and possible. Without his guidance and encouragement, this project would not have been successful.

The submitted manuscript has been authored by a contractor of the U.S. government under Contract DE-AC04-94AL85000. Accordingly, the U.S. government retains a nonexclusive, royalty-free license to publish or reproduce the published form of this contribution, or allow others to do so, for U.S. government purposes.

Part I

Chapter 1

Security Risk Assessment and Management Process

1.1 INTRODUCTION

Since September 11, 2001, decisions for security risk managers have become even more difficult. The terrorist threat potential, that is, the likelihood of an attack, motivations, and capabilities, has dramatically increased. The need to add security features has placed a heavy burden on the already strained budgets of government and commercial enterprises. Some companies have had to decide whether or not they can maintain their business and provide the required security to adequately protect their facilities and the lives of their employees. Security risk managers need a mechanism to help them analyze the information that they do have to make the most logical business decisions to protect their facilities against the very real potential of malevolent acts.

First, managers must define what is essential to the mission of the facility: What are the undesired security events that would interrupt the mission, the consequences associated with the events, the targets that must be protected to prevent the security events, and the liabilities incurred? Concurrent with determining what is important to the mission is identifying what to protect against, that is, defining the adversarial threat spectrum to understand who might attempt the undesired event(s). The adversarial threat spectrum could include international or domestic terrorists, religious or political extremists, criminals, the mentally deranged, or the insider employee. Next, a system effectiveness analysis or vulnerability analysis is completed to determine how well the current security system protects against the adversarial threat spectrum for the undesired events. Once the security system's effectiveness is known, the security risk can be estimated and the manager must assess whether or not the risk level is acceptable. If the risk level is deemed to be too high, the manager must consider the impacts on operations and costs to reduce risk by improving the security system or reducing the consequences. Balancing the resultant impacts and risk reduction can present quite a challenge, but is of utmost importance. (See Figure 1.1.)

This chapter will outline a validated risk assessment and management process that supports managers in determining how much security is enough for their facility, business, or industry. Each following chapter in this book will support one or more steps of the Security Risk Assessment and Management Process. The process can be and has been adapted for various applications, including many elements of our nation's critical infrastructure.

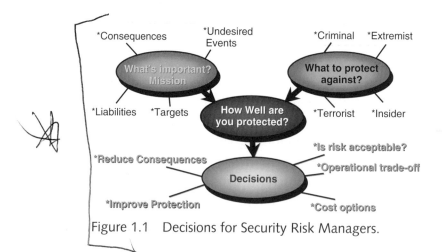

Figure 1.1   Decisions for Security Risk Managers.

The risk assessment and management process was developed at Sandia National Laboratories (SNL) in the 1990s for the Interagency Forum for Infrastructure Protection (IFIP). The IFIP was formed when various related government agencies with common security concerns came together to address security protection against the terrorist threat, as called for by Presidential Decision Directive #63, signed by former President Bill Clinton. Proven physical protection tools and concepts resulting from thirty years of testing and development at SNL were integrated into a single methodology for assessing infrastructure and life-threatening risk. The process was originally applied to the protection of federal dams, high-voltage electric power transmission systems, and other critical national infrastructures. The tool was completed, tested, and published a month before 9/11, and has since been used to estimate relative security risk level and to assess the protection effectiveness and design security and consequence mitigation systems of hundreds of government and commercial facilities against malevolent acts.

However, security risk is difficult to quantify. The traditional risk equation can be used to begin the process. Traditionally, security risk is a function of the likelihood of adversary attack, the likelihood that the adversary attack is successful, and the consequences associated with the loss to the attack. The relative risk estimation process described here is qualitative in nature and allows decision makers to rank events in relative order, to enable them to make risk management decisions. Figure 1.2 describes the three parameters used to estimate security risk.

The conclusions drawn and the information used in the application of the risk assessment process produce sensitive company information that must be protected. The level of protection of the information and the means of protection must be determined, planned, and implemented before the analysis begins. The three factors of the security risk equation each encompass

Figure 1.2   Parameters Used to Estimate Security Risk.

information that, if compromised, could provide serious advantage to the adversary.

1.2   SECURITY RISK EQUATION

Security risk is estimated by the following traditional risk equation:

$$R = P_A * (1 - P_E) * C$$

where:

$$R = \text{risk associated with adversary attack}$$

$$P_A = \text{likelihood of attack}$$

$$P_E = \text{probability that the security system is effective against the attack}$$

$$(1 - P_E) = \text{system ineffectiveness}$$

$$C = \text{consequence of the loss from the attack}$$

Security risk is difficult to quantify, because the basic assumptions for calculating mathematical probability cannot be met; that is, the variables are neither independent nor random. Estimating the likelihood that an adversary will decide to attack a given facility is difficult, at best, because predicting human behavior can never be a random event in the mathematical sense. Humans continually plan, practice, learn, and modify their behaviors. For these reasons, quite often analysts will estimate conditional risk for security applications. Conditional risk presumes that the initiating event occurs (for security applications, this means that the adversary does decide to attack and conducts the attack against the specific facility).

This assumption can focus the risk assessment on the likelihood of adversary success and the associated consequences resulting from the attack. Sometimes building owners and operators need more concrete resolution in risk estimates. They may have several buildings that are vulnerable to the threat, and the consequence of loss is high, but they have credible evidence that makes them believe that one building is more or less likely to be attacked than another, and they feel they must prioritize their security spending, especially if funds are limited.

Various risk assessment and risk management methods have been developed. While each method has its own unique name, focus, and methodology, all attempt to answer three fundamental questions:

1. What are the bad things that can happen to my facility?
2. How likely are the bad things?
3. How do they affect my facility – its mission, occupants, surroundings, and the larger environment?

This text will provide a process to estimate relative security risk based on qualitative estimates for three risk parameters:

- **Likelihood of attack** – Qualitative estimate for likelihood of adversary attack, P_A. Note that threat potential for attack, likelihood of attack, and P_A mean the same thing in this text.
- **Consequence of successful adversary attack** – Qualitative estimate of consequence, C.

- **System ineffectiveness** $(1 - \mathbf{P_E})$ – Qualitative estimate of adversary success or the complement of system effectiveness, P_E.

1.3   SECURITY RISK ASSESSMENT AND MANAGEMENT PROCESS

An analytic process is used to assess security risk. Figure 1.3 describes the order and sequence of the basic steps of the process. The process begins with an optional screening analysis for corporations to prioritize their facilities, followed by characterization of the subject facility, including identification of the undesired events and the respective critical assets. Guidance for defining an adversarial threat is included, as well as for using the definition of the threat to estimate the threat potential for attack or likelihood of adversary attack at a specific facility. Relative values of consequence are estimated. Another optional step allows the owner to prioritize the

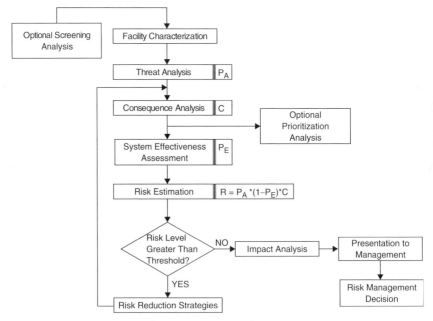

Figure 1.3   Security Risk Assessment and Management Process.

assets at a given facility. Methods are also included for estimating the effectiveness of the security system against the adversary attack. Finally, relative risk is estimated. In the event that the value of risk is deemed to be above a predetermined threshold (too *High*), the methodology addresses a process for identifying and evaluating risk reduction strategies in order to reduce risk.

1.3.1   Facility Characterization

An initial step in security system analysis is to characterize the facility to be analyzed. Facility characterization requires a thorough understanding of the mission and operating conditions of the building, as well as the security concerns. The security concerns should describe the undesired events – the specific events that, ideally, the protection system should prevent. An extension of describing the undesired events is identification of the company's critical assets that an adversary would most likely be attempting to harm or obtain. Sometimes the assets to be protected are obvious by inspection; in complex operations, an analytical logic approach may be required to ensure that all of the critical assets are identified and protected.

Facility characterization includes a complete physical description, not only of the physical layout of the building but also of the construction details, locations of site boundaries, building locations, floor plans, and access points as well as policy and procedures and physical and cyber-protection features and their locations. Any known vulnerabilities or weaknesses in protection are noted.

The facility characterization concludes with a statement of the protection objectives for the facility. Usually, the protection objectives are a list of undesired events or some subset of the undesired events and a listing of the respective critical asset(s) to be protected. For example, a protection objective of a building might be to ensure health and safety for building occupants or to prevent the theft of a particular critical asset.

1.3.2   Threat Analysis

The first parameter of the risk analysis process is the threat potential, particularly, the likelihood of adversary attack.

Threat – Before a vulnerability analysis can be completed and before threat potential for attack or likelihood of attack can be estimated, a description of the threat is required. This description includes the types of possible adversaries, tactics, and capabilities (e.g., number in the group, weapons, equipment, and transportation mode). The threat definition is often reduced to several paragraphs that describe the type and number of adversaries, their modus operandi, the type of tools and weapons they would use, and the type of events or acts they are willing to commit.

The types of organizations that may be contacted during the development of a threat definition include local, state, and federal law enforcement and related intelligence agencies. Local authorities should be able to provide reports on the types of criminal activities occurring and analytical projections of future activities. A review of literature may also be conducted to include past incident reports associated with the site, local periodicals, professional journals, and other related material.

Threat Potential for Attack (Likelihood of Attack) – After the adversarial threat spectrum has been described, the information can be used together with statistics of past events and site-specific perceptions to categorize threats in terms of likelihood that each type of threat would attempt an undesired event. Ideally, the model for security risk assessments could be similar to the model for safety risk assessments; the likelihood of an initiating (abnormal) event is estimated and combined with the likelihood of consequences caused by the initiating event. Safety studies have yielded historical data and statistics that can help predict the likelihood of an abnormal event and the system response to the event. However, estimating the likelihood that an adversary group will

Adversary Capability	Adversary History/Intent	Relative Attractiveness of Asset to Adversary
• Access to region • Material resources • Technical skills • Planning/organizational skills • Financial resources	• Historic interest • Historic attacks • Current interest in site • Current surveillance • Documented threats	• Desired level of consequence • Ideology • Ease of attack

Figure 1.4   Estimating Threat Potential (Likelihood of Attack) for Attack.

attack a specific asset will always represent a challenge because of the human element.

However, a qualitative relative threat potential parameter can be used to estimate the level of the unquantifiable variable. Estimating the threat potential follows a complete threat analysis, and the parameter is estimated per undesired event and per adversary group. The basis of the parameter estimation is:

- Characteristics of the adversary group relative to the asset to be protected
- Relative attractiveness of the asset to the adversary group

Figure 1.4 includes information that can be used to estimate the likelihood that a given adversary group would decide to attack a specific facility.

1.3.3   Consequence Analysis

The second parameter of security risk is consequence. Consequence analysis can be completed after the undesired events and associated critical assets have been identified as a part of facility characterization. The next analysis step is to estimate consequences associated with the loss of specific critical asset(s) for each undesired event. Consequence definitions are site- or industry-specific. Organizations describe consequence in categories or terms

Table 1.1   Consequence Definitions

Consequence Category	Consequence Level
Could result in death, permanent total disability, loss exceeding $1M, or irreversible severe environmental damage that violates law or regulation.	Catastrophic
Could result in permanent partial disability, injuries, or occupational illness that may result in hospitalization of at least three personnel, loss exceeding $200K but less than $1M, or reversible environmental damage causing a violation of law or regulation.	Critical
Could result in injury or occupational illness resulting in one or more lost workday(s), loss exceeding $10K but less than $200K, or mitigatible environmental damage without violation of law or regulation, where restoration activities can be accomplished.	Marginal
Could result in injury or illness not resulting in a lost workday, loss exceeding $2K but less than $10K, or minimal environmental damage not violating law or regulation.	Negligible

that are meaningful to them; some may measure consequence in terms of lost income or downtime, others in casualties or illness, and others in terms of loss of pubic confidence or reputation. The consequence categories, such as dollars, deaths, injuries, downtime duration, and negative publicity, that characterize consequence must be determined first. Further, definitions must be established for qualitative levels for each consequence category. Table 1.1 provides an example of a Consequence Definition Table that is similar to one used by the Department of Defense in accordance with Military Standard 882D. The primary goal of consequence analysis is to estimate the relative consequence value associated with each undesired event due to loss or compromise of a critical asset.

1.3.4   System Effectiveness Assessment

The third parameter in assessing security risk, system ineffectiveness $(1 - P_E)$, can be derived from a security system effectiveness assessment. Security system ineffectiveness (adversary success) and security system effectiveness (P_E) are complementary functions. If security system effectiveness is *High*, then security system ineffectiveness (adversary success) is judged to be *Low*. The risk assessment process will evaluate security system effectiveness in order to estimate system ineffectiveness (adversary success). A defensible measure of the effectiveness of the security system to prevent the undesired events for the given threat spectrum is an important factor in the security risk equation.

The process focuses on security system effectiveness assessment. A valuable product of assessing system effectiveness is the identification of specific vulnerabilities of the protection system. If the security system effectiveness is judged to be *Low*, specific weaknesses and the associated deficient protection elements causing the *Low* level are site-specific system vulnerabilities. Knowledge of site-specific vulnerabilities is valuable for planning system upgrades to reduce risk and for contingency planning to know where to place reinforcement protection during times of elevated threat conditions.

For most applications, a security system is made up of physical protection features and cyber-protection features. Some undesired events can be accomplished by a physical attack on the facility, whereas others can be accomplished by a cyber-attack on the system. A total security system should address both physical and cyber-attacks, as appropriate. A complete system effectiveness assessment will include a physical protection analysis and cyber-protection analysis.

1.3.4.1 Physical Protection System Effectiveness

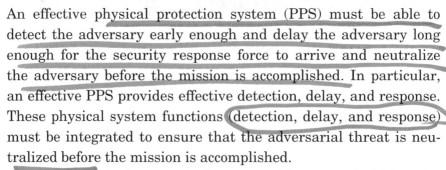

An effective physical protection system (PPS) must be able to detect the adversary early enough and delay the adversary long enough for the security response force to arrive and neutralize the adversary before the mission is accomplished. In particular, an effective PPS provides effective detection, delay, and response. These physical system functions (detection, delay, and response) must be integrated to ensure that the adversarial threat is neutralized before the mission is accomplished.

DETECTION, the first required sequential function of a PPS, is the discovery of adversary covert or overt actions and includes sensing actions. In order to discover an adversary action, the following events must occur:

- Sensor (equipment or personnel) reacts to an abnormal occurrence and initiates an alarm
- Information from the sensor and assessment subsystems is reported and displayed
- Someone assesses the information and determines the alarm to be valid or invalid

DELAY is the second required function of a PPS. Any feature that impedes adversary progress can be considered to be delay. Delay can be accomplished by barriers (e.g., doors, vaults, locks) or by distances that cause a time delay to traverse. The security protective force can be considered an element of delay if personnel are in fixed and well-protected positions.

RESPONSE, the third requirement of a PPS, comprises actions taken by the security police force (law enforcement officers) to prevent adversarial success. Response consists of interruption of and neutralization of the adversary action.

1.3.4.2 Cyber-Protection System Effectiveness

Much like an effective PPS that demonstrates high performance for the three functions of detection, delay, response, and the integration of these functions, an effective cyber-protection system demonstrates high performance for three basic cyber-security functions and their integration. These functions are used to ensure the properties of confidentiality, integrity, and availability of data. *Confidentiality* requires that information not be made available to unauthorized individuals, entities, or processes. *Integrity* requires that information not be altered or destroyed in an unauthorized manner. *Availability* requires that information be accessible and usable on demand by an authorized entity. The three cyber-protection functions are:

- Authentication
- Authorization
- Audit

The authentication, authorization, and audit must be performed at a high level and must be integrated. The authentication and authorization strategies both provide data to the audit capability where it is analyzed for evidence of malicious activity.

Authentication – Authentication establishes the validity of a claimed identity. User authentication is the capability of associating a computer identity with a human being. This may be done using mechanisms that fall into three categories: (1) something the individual knows, (2) something the individual has, and/or (3) something the individual is. Once a user is authenticated, he or she is generally issued credentials that are associated with computer processes acting in the user's behalf. User authentication is critical to the overall security of a system or network, because if one user obtains (maliciously or otherwise) another user's credentials, then he or she can access any information that the user is permitted to access.

Authorization – Authorization determines what actions an entity is allowed to perform with respect to a given information object (e.g., files, database records, web pages). Authorization for access to systems and applications must be granted by management. Authorization for access to information on systems must be controlled so that only authorized users can access specified information objects, based upon their authenticated identity.

Audit – Audit records the actions or attempted actions performed by an entity within a computer system or network. The cyber-intrusion detection system supports the audit function. The major components of a successful cyber-intrusion detection system are the continual review of traffic data, scanners that detect any unusual occurrences, including any suspect ports or modems, virus protection, and monitors for access control.

Access control monitoring ensures a complementary relationship between firewalls and intrusion detection systems. Firewalls block undesired network traffic and permit desired traffic. The cyber-intrusion detection system inspects both blocked and permitted traffic for suspect patterns.

1.3.4.3   Security System Performance Assessment

Analysis and evaluation of protection systems begins with a review and thorough understanding of the protection objectives and security environment. Analysis can be performed by simply checking for the required features of an effective protection system, such as intrusion detection, entry control, access delay, response communications, and a response force for a physical system and features for authentication, authorization, and audit for a cyber-protection system. However, a system based on required features usually does not lead to a high-performance system because those features are often not integrated to ensure adequate levels of protection for the identified threat spectrum. Sophisticated analysis and evaluation techniques can be used to estimate the minimum performance levels achieved by a security system. The most reliable effectiveness measure is performance as a total integrated system.

1.3.5   Risk Estimation

Security risk is a function of the likelihood of attack, consequence of successful attack, and security system ineffectiveness. To estimate relative security risk, the qualitative estimates for likelihood of attack, system ineffectiveness, and consequence are logically combined. A simple method, based on expert judgment, for combining the three risk parameters to estimate security risk will be discussed. The security risk estimates are relative, not absolute, but they can be used to make risk management decisions. A relative risk level is valuable to:

- Compare risk levels for a spectrum of malevolent threats
- Compare risk levels for a spectrum of facilities, industries, or organizations
- Compare the cost-effectiveness and other impacts of potential improvements

1.3.6   Comparison of Estimated Risk Levels

Estimated risk levels are compared to a predetermined risk threshold to decide whether further analysis is required. The threshold is determined by the analysis team and the security risk managers.

1.3.7   Risk Reduction Strategies

If the estimated baseline risk level for the threat spectrum is judged to be above the established threshold (too *High*), risk reduction strategies for the system may be considered. Risk reduction strategies focus on reducing the levels of the parameters of the security risk equation: likelihood of attack, system ineffectiveness, and consequence. In practice, risk reduction is made most successful by improving protection system effectiveness and mitigating consequences.

Risk Reduction Upgrades – Security system planners must address how to reduce security risk. Planners might consider adding features to increase physical or cyber-protection system effectiveness and/or to reduce or mitigate consequences. Site-specific vulnerabilities identified in the system effectiveness analysis provide guidance for adding/modifying features. Upgrades to the system might include retrofits, additional safeguard features, or additional consequence mitigation features. Consequence analysis and system effectiveness analysis should then be repeated for the upgraded system in order to estimate a risk level associated with the upgraded system. If the estimated risk for the upgraded system is below the threshold, the upgrade is completed. If the risk is still above the threshold, the upgrade process should be repeated until the risk level is judged to be below the threshold.

Impact Analysis – Once the system upgrade has been determined, it is important to evaluate the impacts of the risk reduction on the mission of the facility and the cost. If system upgrades put a heavy burden on normal operation, a trade-off would have to be considered between risk and operations. Budget can be the driver in implementing security upgrades. A trade-off between risk and total cost may have to be considered. The assessed level of risk and the upgrade impact on cost, mission, and schedule are valuable information to security risk managers.

1.4   PRESENTATION TO MANAGEMENT

The final step in the risk assessment process is the preparation of a presentation package for the risk managers and stakeholders. The presentation generally includes the threat description, the security risk estimates for the baseline system, descriptions of any risk reduction packages, and the results of the impact analysis for the risk reduction package(s). By using comparison to the baseline risk levels, managers are able to understand what the upgrade package is buying them in risk reduction as well as other potential impacts. The total presentation package provides invaluable information for risk management decision makers.

1.5   RISK MANAGEMENT DECISIONS

Building owners, stakeholders, and risk managers have the risk assessment information package to help them make difficult security decisions. Most importantly, risk managers must decide on the design basis threat or the threat level to which the security system will be designed.

1.6   INFORMATION PROTECTION

The risk assessment process provides valuable, detailed information for risk managers; likewise, the information could provide valuable information to any potential adversaries. Because the process begins with basic facts and assumptions and each step builds on previous step(s), allowing the information to get into the wrong hands could provide a roadmap for the malevolent threat. Each step of the process provides security sensitive information:

1. **Facility characterization** identifies the security concerns, critical asset(s), and their locations.
2. **Threat analysis** ultimately defines the level of protection to which the security system is designed. If the perceived highest threat level is the terrorist, the security system will be designed to be much stronger than if the perceived threat is the vandal.
3. **Consequence analysis** prioritizes the assets in terms of criticality or value.
4. **System effectiveness assessment** provides possible attack scenarios and documented system weaknesses or vulnerabilities.

For these reasons, once the process is applied to a specific facility, the entire analysis package must be protected. Most sites will have to develop the infrastructure for protecting, storing, and sharing the risk assessment package.

1.7   PROCESS SUMMARY

This chapter provides an overview of an analytical security risk assessment and management process. Application of the risk assessment process supports managers in determining how much security is enough for their facility, business, or industry. The required steps of the process are:

1. Characterize the facility.
2. Analyze the malevolent threat and estimate the threat potential for attack of the facility.
3. Estimate consequences associated with the attack.
4. Assess the effectiveness of the physical and cyber-protection systems.
5. Estimate relative security risk as a function of likelihood of attack, security system ineffectiveness, and consequence.
6. Compare the security risk level to a predetermined threshold.
7. Suggest risk reduction strategies if the estimated risk level is above threshold, followed by re-evaluating consequences and protection system effectiveness to measure and ensure relative risk reduction.
8. Analyze impacts imposed by risk reduction packages.
9. Present completed assessment to management.
10. Make risk management decisions.

Risk Assessment Process

The process begins with basic facts and assumptions, and each step builds on previous step(s). The final results are defendable because they are traceable back to the original facts and assumptions. Results are repeatable, and updates to any step are easily addressed without starting all over. The process can be adapted to assess the security risk for most entities. The security of dams, energy infrastructures, chemical facilities, buildings, and communities has been enhanced by the application of the process.

1.8   REFERENCES

1. Biringer, Betty, *Risk Assessment Method for Electric Power Transmission,* presented at Carnahan Conference on Security Technology, sponsored by IEEE, Albuquerque, NM, October 2004.

2. MIL-STD-882D, *Department of Defense Standard Practice for System Safety*, February 10, 2000.
3. *North American Electric Reliability Council, Urgent Action Cyber Security Standard,* Standard CIP-002-1, Draft, May 9, 2005, http://www.nerc.com/~filez/standards/Cyber-Security-Permanent. html.
4. *Sandia National Laboratories Security Risk Assessment Methodologies*, http://www.sandia.gov/ram.
5. Wyss, Gregory, D., "Risk Assessment and Risk Management for Energy Applications," in *Energy 2000: State of the Art*, ed. Peter Catania, Balaban Publishers, L'Aguila, Italy, pp. 163–184, 2000.

1.9   EXERCISES

1. Of what value is a security risk assessment to security risk managers? Justify your answer.
2. List and describe the parameters used to estimate security risk.
   a. Are these parameters mathematically independent? Why or why not?
   b. Can these parameters be quantified? Why or why not?
   c. Must these parameters be estimated in any given order? Why or why not?
3. Discuss estimating the threat potential for attack:
   a. What are the limitations, if any?
   b. What are important considerations?
4. Discuss estimating security system effectiveness.
   a. Why is it important to consider both physical protection system effectiveness and cyber-protection system effectiveness?
   b. Discuss the relationship between security system effectiveness and adversary success.
5. Discuss estimating the consequences of adversary attack.
   a. What are some possible parameters to define or describe consequence?
   b. What are consequence mitigation features? Define and provide examples.
6. What choices do managers have if security risk level is deemed to be too *High*? Describe ways to reduce security risk.

7. Why is it important to consider all impacts when considering security system upgrades?

8. How are safety and security risk assessments alike? How are they different?

9. How might the results of a security risk assessment be used for security contingency planning? Security contingency planning describes procedures or features that are implemented during elevated threat conditions for events that are otherwise very *Low* likelihood but *High* consequence.

10. How might potential adversaries use either input information or results of the security risk assessment for a given site?

Chapter 2

Screening Analysis

2.1 INTRODUCTION

Screening analysis is an optional step before an investment is made in a complete security risk assessment. Complete security risk assessments represent a commitment of time and resources. Sometimes owners of multiple facilities have limited time, staff, and/or dollars to invest in security, so they first want to know which of their facilities warrant a full risk assessment. For those that require an assessment, it might be helpful to know which (based on security risk) should be done first, if there are multiple facilities. If resources only allow one or a limited number of analyses to be completed, which facility should be analyzed? A screening analysis will help analysts and decision makers prioritize facilities, if necessary.

2.2 SCREENING ANALYSIS METHODS

If a security screening analysis is judged to be necessary, the complexity and depth of the screening model depends on the security concerns, the number of facilities to be screened, and the variation among the facilities. If the list of security concerns is not consistent among the sites, if there are numerous facilities to be screened (hundreds to thousands), or if the sites do not vary much, analysis methods must be more complex to provide the needed differentiation for screening. Usually, for security screening, a single parameter is used, and the consequence impact associated with the loss is the preferred parameter.

Before the screening analysis can begin, a standard must be established for addressing consequence level. (Later in Chapter 5, "Consequence Analysis," a Reference Table of Consequences will be developed that is much more detailed and specific than the standard suggested here for screening. However, if the analysis team is willing to invest the effort upfront, the Reference Table of Consequences could be developed first and used for the screening analysis.) The standard should be developed by the analysis team with approval by management. The list of criteria for measuring or describing consequence level should be specified as well as definitions for qualitative levels of *High*, *Medium*, and *Low* for each criterion. Common criteria used for comparing consequence level include number of deaths, economic loss, loss of production, and/or downtime. Further, comparisons should be made to ensure that the levels for the criteria are consistent. For example, the *High* estimate should represent the same level of consequences for all criteria; the *Medium* estimate should represent the same level of consequences for all criteria; the *Low* estimate should represent the same level of consequences for all criteria.

Figure 2.1 describes the steps of a security screening method based on consequence.

First, facilities must be reviewed. For each facility, a consistent set of consequence categories must be estimated for each security concern or undesired event, if it should occur at the facility. Undesired events are those security concerns or malevolent acts that the security system is trying to prevent. Examples might be the theft of high-value items, the destruction of a critical asset that interrupts the mission of the facility, or harm to the occupants of the building. Judgments for levels must be kept as consistent as possible and should be described as *High*, *Medium*, or *Low* consequence

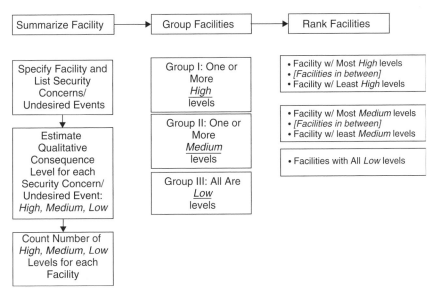

Figure 2.1   Security Screening Analysis Method.

impact. The maximum impact for each undesired event should be recorded. Then each facility can be summarized with a tabulation of the number of *High*, *Medium*, or *Low* levels estimated for each undesired event. The list of facilities can then be divided by grouping all facilities with one or more *High* levels in Group I, those facilities with one or more *Medium* levels in Group II, and finally, those facilities with only *Low* levels in Group III. Further, the facilities can be ranked by organizing each group in terms of the facility with the most undesired events at the given level to the facility with the least number at the given level.

Consider this example: Alpha Corporation owns and operates seven multistory office buildings in several large cities in the Southwest. Various private and government organizations lease space in the buildings. The security management team at Alpha has decided that they want to improve the security level of their buildings. The consequence impacts for the selected undesired events and the consequence factors that are common to all buildings are estimated and are summarized in the Table 2.1.

Table 2.1   Example Screening Analysis

Building	Undesired Event	Aircraft Impact into Building	Vehicle Bomb	Theft of Valuable Assets	Briefcase Bomb / Suicide Bomber
A	Loss of building structure (economic)	M	L	M	L
	Loss of operations (economic)	M	L	L	L
	Loss of lives	L	L	L	L
	Environmental impact	L	L	L	L
	Maximum impact:	M	L	M	L
B	Loss of building structure (economic)	H	M	L	M
	Loss of operations (economic)	M	L	L	L
	Loss of lives	H	H	L	H
	Environmental impact	L	L	L	L
	Maximum impact:	H	H	L	H
C	Loss of building structure (economic)	H	L	L	L
	Loss of operations (economic)	M	M	L	M
	Loss of lives	H	H	L	H

Table 2.1 (continued)

Building	Undesired Event	Aircraft Impact into Building	Vehicle Bomb	Theft of Valuable Assets	Briefcase Bomb / Suicide Bomber
	Environmental impact	L	L	L	L
	Maximum impact:	H	H	L	H
D	Loss of building structure (economic)	M	L	M	L
	Loss of operations (economic)	H	M	L	M
	Loss of lives	H	H	L	H
	Environmental impact	L	L	L	L
	Maximum impact:	H	H	M	H
E	Loss of building structure (economic)	M	L	L	L
	Loss of operations (economic)	L	L	L	L
	Loss of lives	L	L	L	L
	Environmental impact	L	L	L	L
	Maximum impact:	M	L	L	L

Table 2.1 (continued)

Building	Undesired Event	Aircraft Impact into Building	Vehicle Bomb	Theft of Valuable Assets	Briefcase Bomb/ Suicide Bomber
F	Loss of building structure (economic)	M	M	L	M
	Loss of operations (economic)	M	L	L	L
	Loss of lives	M	L	L	L
	Environmental impact	L	L	L	L
	Maximum impact:	M	M	L	M
G	Loss of building structure (economic)	L	L	L	L
	Loss of operations (economic)	L	L	L	L
	Loss of lives	L	L	L	L
	Environmental impact	L	L	L	L
	Maximum impact:	L	L	L	L

The table completes the first step of the screening analysis, which is to summarize the facilities in terms of the consequence impacts for each undesired event for each facility. The next step is to group the facilities. The groups are:

Group I (one or more *High* consequence impact levels):

- Building B
- Building C
- Building D

Group II (one or more *Medium* consequence impact levels):

- Building A
- Building E
- Building F

Group III (all are *Low* consequence impact levels):

- Building G

The final step of the screening method is to rank the facilities in order of consequence severity:

1. Building D (3 High, 1 Medium)
2. Building B and Building C (3 High, 1 Low)
3. Building F (3 Medium, 1 Low)
4. Building A (2 Medium, 2 Low)
5. Building E (1 Medium, 3 Low)
6. Building G (4 Low)

The security risk management team at the Alpha Corporation could use the results of the screening analyses for planning purposes. If resources, time, and/or money only allow them to complete a few full security risk assessments for the year, Buildings D, B, and C would be good choices, based on consequence if attacked. There is danger if the screening analysis is used to delete facilities from the risk assessment list; screening analyses should be used to prioritize the facilities and to schedule the security risk assessments for the facilities.

2.3   SUMMARY

A simple method for screening analysis has been presented. The method is based on consequence impact. If this screening method does not provide enough discrimination, the consequence impact definitions can be redefined to provide more levels of differentiation, for example *very low, low, medium, high,* and *very high*. Another option is that an additional parameter can be used to provide more refinement. Chapter 6, "Asset Prioritization," will discuss a prioritization scheme that is based on both a consequence impact as well as a threat potential parameter.

2.4   REFERENCES

1. Biringer, Betty, "Risk Assessment Method for Electric Power Transmission," presented at Carnahan Conference on Security Technology, sponsored by IEEE, Albuquerque, NM, October 2004.
2. Matalucci, Rudy and Strothman, John, "Security Risk Assessment Procedures: Countering Terrorism and Other Threats," Infrastructure Security Course sponsored by ASCE and Sandia National Laboratories, Las Vegas, NV, January 26–27, 2006.

2.5   EXERCISES

1. Describe circumstances under which security risk managers would benefit from a screening analysis.
2. What parameters and definitions in a screening analysis must be kept consistent and why?
3. Discuss how the results of a screening analysis could be misused.
4. Suppose that you were the owner or manager of a business that had 500 buildings located across the country, and security alert conditions became elevated to a point that security managers decided that security effectiveness needed to be reviewed at their facilities. Your security team completes an initial screening analysis based on consequence and estimates that about 300 of the buildings rank at about the same relative (high) level and the remaining 200 are far below this. Discuss several options for proceeding, and list the advantages and disadvantages of each option.

Chapter 3

Facility Characterization

3.1 INTRODUCTION

Before a specific site security risk assessment can be launched, a complete characterization of the facility must be completed. The total environment must be captured: the security environment, the operations environment, and the workforce environment. Assessment team members from the site will know or be able to acquire the information readily. If the assessment is completed by someone other than site employees, collection of all of the information can be a substantial task. Site or construction drawings, maps, safety reports, site surveys, process descriptions, tours, and interviews are sources of information.

The facility characterization provides the foundation for the risk assessment. The essential products of a complete characterization are shown in Figure 3.1 and include:

- Identification of all of the undesired events
- Description of the facility
- Identification of the critical assets requiring protection
- Specification of the protection objectives for the security system

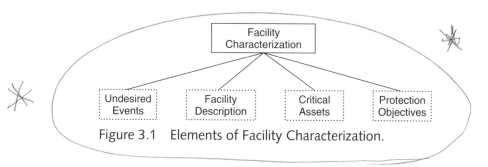

Figure 3.1   Elements of Facility Characterization.

3.2   UNDESIRED EVENTS

The initial task of facility characterization is to determine the security concerns at the specific site. Security concerns range from those that result in catastrophic consequences to those that are primarily nuisances or embarrassments. With input from management, the assessment team should identify a list of undesired events. Management is usually willing to consider expending resources to keep specific undesired events from occurring. For a security risk assessment, usually a more complete list of undesired events is used than that of the screening analysis.

Undesired events are facility-dependent and are generally associated with loss of mission or threats to public health and safety. Examples of undesired events for a building are:

- Disruption of operations
- Theft of valuable assets
- Crimes against people
- Destruction of the building
- Compromise of the information management system
- Loss of public confidence

○ SERVERS
○ TRAIN Derailment

Disruption of operations could involve various events, including sabotage of physical equipment or support systems such as the emergency system, power, or other utilities. Theft of valuable assets could be theft of assets that are important to the mission or impose a large economic impact. Crime against people could be bombs, snipers, a chemical or biological attack, or a kidnap or hostage situation. Undesired events concerning destruction of the building could be due to vehicle or suitcase bombs or a result of aircraft impact. Compromising the information management system would include a cyber-attack on information systems associated with operations or safety systems. Loss of public confidence could result from any incident that causes embarrassment or destroys the reputation of the facility or corporation.

3.3 FACILITY DESCRIPTION

Collecting data to adequately describe a facility in order to complete a security risk assessment is of utmost importance and can be one of the more time-consuming tasks. A complete description of the facility and its operating environment is necessary. Data can be collected by team members from site reports (environmental impact statements, safety analyses), construction drawings, and site visits, including tours and interviews with personnel. The type of information to be collected in order to adequately describe the facility is:

- Physical details
- Cyber-information-system details including cyber-protection features
- Facility operations
- Security protection systems
- Safety protection systems
- Workforce description
- Restrictions, requirements, limitations

3.3.1 Physical Details

A physical description of the facility provides details on the boundary and all of the penetrations of the boundary as well as the topography and landscape of the area. The building should be described, including construction details, closest vehicle parking distances, entrances and their locations, utilities, the types and locations of utility penetrations, room layout, door construction and lock description, location of fresh-air intakes, design of the heating and ventilation/cooling system, and location of any hazardous materials or waste products. Any layout or structural weakness of the site that reveals situations or conditions that an adversarial threat could use to enhance their chances of success should be specifically noted.

The adversarial threat to the facility is important information to the assessment. In practice, threat analysis may be conducted at the same time as facility characterization. The next chapter will discuss this data-gathering and threat assessment process in great detail.

3.3.2   Cyber-Information System

The architecture of the information system must be understood and documented in detail. The information system discussion will be limited to information systems associated with supervisory control and data acquisition (SCADA) and process control systems. All of the access points to the system must be documented, located, and described in detail. Some facilities may have a system expert on-site; others may contract out the design, maintenance, and operation of their information system. The level of information needed to adequately describe the information can usually only be obtained from the technical expert who operates and maintains the system.

3.3.3   Facility Operations

The facility operations required to accomplish the mission of the facility must be identified, located, and described. Any appropriate policy and procedures must also be noted. The operating environment must be understood; the working hours/off hours, the required processes for normal operations, as well as changes that occur during emergency, construction, or other conditions must be addressed, especially if the protection system changes during different operating states. Any conditions or situations that an adversarial threat could exploit to enhance their chances of success should be highlighted.

3.3.4 Security Protection Systems

Both the physical protection system (PPS) and the cyber-protection system must be described in terms of function, features, location and description of features, as well as any known weaknesses or gaps in protection.

PPS – The PPS is made up of features that support the detection, delay, and response functions.

Detection, the first required function of a security system, is the discovery of adversarial action and includes sensing a covert or overt action. The components of detection include some type of sensor (either equipment or personnel) that initiates an alarm, communication of alarm information, and assessment of information to determine if the alarm is valid or invalid.

Methods of detection include a wide range of technology and personnel. Entry control, a means of allowing entry of authorized personnel and detecting the attempted entry of unauthorized personnel and contraband, is included in the detection function. Entry control, where it includes locks, may also be considered a delay factor in some cases. Searching for metal (possible weapons or tools) and explosives (possible bombs or breaching charges) is required for some high-security areas. This may be accomplished using metal detectors, x-raying (for packages), and explosive detectors. Security officers or other personnel can accomplish detection if they are trained in security concerns and they have the means to alert the authorities in the event of a security problem.

Applicable detection information would include documentation of all sensors, exterior (fence, motion, door, or gate) or interior (door, motion, proximity). Equally important is notation of how the sensors operate in their environment, that is, are they the appropriate sensors for the application; are they installed, tested, and maintained properly; and do they perform as expected? Also, are the sensors adequately protected so that they cannot be easily tampered with or bypassed? The security and reliability of communication of alarms generated by sensors or personnel are important. The existence of a designated location where alarms can be received and assessed should be confirmed. Locations of personnel, work hours, and whether or not personnel have a reliable means to communicate a problem or duress should be recorded. Any entry control features, including badges, contraband checks, locks, and the policy and procedures associated with them should be noted.

Any known gaps or problems with the detection features should be specifically noted in the facility characterization process. Mary Lynn Garcia in *Vulnerability Assessment of Physical Protection Systems* provides detailed guidance on how to identify specific weaknesses in the detection function.

Delay is the second required function of a physical security system. Features that impede adversary progress can contribute to the delay function. Delay can be accomplished by fixed barriers (e.g., doors, vaults, locks) or traversal distances. In high-security applications, delay is sometimes achieved by sensor-activated barriers, such as dispensed liquids, smoke, and foams or a protected, armed security personnel force. For most security applications, nonsecurity personnel are not considered a delay feature. The response force may serve as a delay if they arrive in time to intercept the adversary before the undesired event is caused.

Traversal distances and locks are also considered delay features. Site layout and construction drawings can be used to document building traversal distances and wall, roof, and door construction. Lock type (key, code, mechanical, or electronic) as well as the procedures for who has keys or combinations should be described.

The conditions under which locks and combinations are changed should also be considered.

Response, the third requirement of physical security systems, comprises actions taken by the security police force (law enforcement) to prevent adversarial success. Most sites are not equipped or legally able to have an on-site personnel protective force. The response force has to arrive in time to interrupt the adversarial action while it is still in progress. How the site communicates with local law enforcement and how reliable and timely the response could arrive should be established. Whether or not the site has specific agreements with local law enforcement is important.

Cyber-protection system – The architecture of the protection system for the SCADA or process control system must be described in significant detail. All of the access points to the system must be identified. Systems can be accessed via modems (located on or off-site), the Internet, control room, alternate access points in the facility, communication links, or by the download of software. Any electronic security boundaries of the system must be described. Normally, cyber-systems have an exterior electronic boundary and one or more interior boundaries. The communication links between boundaries must be identified. Cyber-protection features are usually deployed at these electronic boundaries. The cyber-protection system is made up of features that support the authentication, authorization, and audit functions.

Authentication is the process of establishing the validity of a claimed identity. User authentication is the process of associating a computer identity with a human being. This may be done using mechanisms that fall into three basic categories: (1) something the individual knows, (2) something the individual has, and/or (3) something the individual is. Two-factor authentication means authentication requiring two (or more) of the above factors. Whatever the mechanism is used for authentication, the policy/procedures and practice of deriving and implementing them should be described in detail.

Authorization is the process of determining what actions an entity is allowed to perform with respect to a given object. Authorization for access to information on systems must be controlled so that only authorized users can access specified information objects (e.g., files, database records, web pages) based upon their authenticated identity. A note should be made if all employees are granted the same access level by management or if access is compartmentalized. Compartmentalization means that some compartments of the cyber-system require a higher access level than others; all employees do not have authorized access to all compartments. Because of their role in cyber-security, all process control network authentication servers are usually afforded the maximum protection that is practical. In order to maintain the confidentiality and integrity of these central authentication services, the number of persons with privileged access (e.g., root or administrator) to these services is kept to a minimum.

Audit is the process of recording the actions or attempted actions performed by an entity within a computer system or network. The intrusion detection system supports the audit function. The major components of a cyber-intrusion detection system include the review of traffic data; scanners to detect any unusual occurrences, including any suspect ports or modems; virus protection; and monitors for access control. Audit features, their procedures and implementation in the cyber-protection system should be described in as much detail as possible.

Any gaps, weaknesses, or absence of features in the cyber-protection system should be specifically noted in the facility characterization process.

3.3.5 Workforce Description

A description of the workforce is important to understanding the security environment at a facility. A listing of the types of positions, as well as the authorization that is afforded to each position in terms of unrestricted access to the site, key operations, and information systems. Typical types of positions are:

- Managers
- Operators/technicians
- Security personnel
- Administrators
- Custodians
- Maintenance personnel
- Contractors
- Vendors
- Visitors

Any pre-employment background investigation programs should be described. Some corporations complete a check with local law enforcement, credit institutions, or past employers before hiring. Special note should be made if investigations are only completed once, before employment, or if periodic re-evaluations are conducted after employment begins. Some institutions conducting national-security-level projects might require a government-issued security clearance and/or certification in a human reliability program that impose very stringent background checks and personnel screening prior to employment.

The general work environment should be evaluated and understood. Any past or present issues with labor relations should be noted. Any past insider incidents of disgruntlement, violence, or crime should be included in the workforce description.

3.3.6   Restrictions, Requirements, Limitations

Some facilities and/or industries are subject to compliance requirements by the government, owner, or operator. The terms of legal requirements or limitations should be described. Note of limitations or restrictions imposed by process operations will be helpful in considering protection system upgrades, if needed. Examples include environmental regulations, safety requirements, operational limitations, or labor union requirements.

3.4   CRITICAL ASSETS

Once the undesired events have been specified and the facility described, the next step in facility characterization is to determine what specific assets must be protected to prevent the undesired events from occurring. The assets that must be protected in order to prevent the undesired event are labeled the critical assets. For some applications, identification of the critical assets can be done by simple observation or inspection. For example, in a jewelry manufacturing facility, if the undesired event is to prevent the theft of precious gems, the gems are the obvious critical assets. In more complex systems, critical assets may not be so obvious, and a logic diagram may be required to identify all of the ways that the undesired events can occur and what assets must be protected to prevent the undesired events from occurring.

3.4.1   Generic Fault Tree

A fault tree is a logic diagram. The fault tree graphically represents the components and subsystems of events that can result in a specified undesired event. Identification and evaluation of the specific assets that make up the components and subsystems that are important to the occurrence of the undesired event leads to the identification of the critical assets that must be protected in order to prevent the undesired event. "Appendix A: Generic Fault Tree for Buildings" contains definitions for fault tree terms and a generic fault tree for various undesired events associated with buildings.

Two kinds of logic gates are used in the generic tree for a building, the **AND** gate and the **OR** gate. Gates have inputs and may or may not have an output. Inputs enter the bottom of the gate; outputs exit the top of the description rectangle above the gate.

The shape of the **AND** gate is a round arch with a flat bottom.

For the undesired event described above the **AND** gate to occur, all of the events that input into the **AND** gate must occur. Thus, if any one of the input events can be prevented, the event described above the **AND** gate will be prevented.

The shape of the **OR** gate is a pointed arch with a curved bottom.

For the undesired event described above the **OR** gate to occur, any one (or more) of the events that input to the **OR** gate must occur. All of the input events must be prevented in order to prevent the event described above the **OR** gate.

The TRANSFER operation is represented by an upright triangle.

The transfer operation is used to make the graphic display of the logic tree more compact and readable or to develop common logic only once. Because many logic diagrams, as they are developed, occupy a wide left-to-right space across the page, it might be necessary to disconnect the development of an event and place it at a more convenient position on the page or on another page. To connect the event and its development without drawing a line between separate figures, the transfer symbol is used. In another use of the transfer operation, the same event or tree branch may apply more than one place on the tree; the event will be developed once, and the transfer symbol will be used to delineate all of the places on the tree that the branch feeds into. The number inside the triangle identifies the logic development.

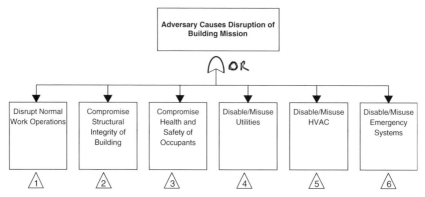

Figure 3.2   Tree Top for Generic Fault Tree for Building.

Figure 3.2 provides the top level of the generic tree for buildings for the adversary causing a disruption of building mission. Note the OR gate at the top; the adversary can cause disruption of building mission by (1) disrupting normal work operations, *or* (2) compromising the structural integrity of the building, *or* (3) compromising the health and safety of occupants, *or* (4) disabling or misusing the utilities, *or* (5) disabling or misusing the heating, ventilation, and air conditioning (HVAC), *or* (6) disabling or misusing the emergency systems. The transfer symbols below each event rectangle identify the logic development that fits below each undesired event. An example of the logic development for each of the tree branches numbered 1 through 6 is provided in "Appendix A: Generic Fault Tree for Buildings." An informal fault tree can be used as a valuable aid to the general thought process when discovering and enumerating critical assets. A more formal fault tree can be used to identify the root causes for undesired events in complex systems. More information on fault tree analysis can be found in the *Fault Tree Handbook* published by the U.S. Nuclear Regulatory Commission.

3.4.2   Identifying Critical Assets

The generic fault tree can be modified to describe a particular facility. The modification process includes removing the tree branches that do not apply, adding any that have not been included in the generic tree, and developing all branches of the tree to the extent that the specific critical assets that must be protected in order to prevent the undesired event can be specified. The completed tree that has been modified for a facility can be used not only to identify the critical assets that must be protected but also to outline strategies or even scenarios for causing the undesired events. Because of the sensitivity of the information provided by the site-specific fault tree, the tree should be well protected and controlled because it would provide sensitive information to an adversary if it got into the wrong hands.

The tree branch for electric power can be used to demonstrate some of the concepts that have been discussed. Electric power is required for various operations at an example site and so logically it is an input to several places on the generic fault tree for building mission. The branch of the tree that develops the logic for electric power at an example building can be developed once and then the transfer symbol can be used to show all of the places where the tree branch for electric power would fit (Figure 3.3).

Note the **AND** gate indicated on the first level: the adversary would have to defeat the commercial power system *and* the emergency generator system *and* the uninterrupted power supply (UPS) batteries in order to eliminate the electric power at the example site. The generic tree can be made site-specific for the example facility by eliminating features that are not present and further developing existing features. In Figure 3.2, the dashed lines represent the additions to the generic building tree to make it site-specific for an example facility. Locations can be identified for specific features, such as the transformer vaults and the substation for the commercial power system; the emergency generator system can be further described in terms of the adversary defeating the diesel fuel system (either the storage tank *or* the fuel lines) *or* the cooling system *or* the auto start system.

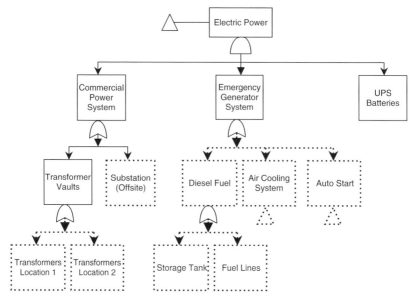

Figure 3.3   Modification of Generic Tree for Example.

If electric power is required for the example building mission, some of the critical assets to be protected would be the assets described at or near the bottom of the tree, namely, the transformers, the substation, the diesel storage tank, diesel fuel lines, components of the cooling system, and components of the autostart system.

3.5   PROTECTION OBJECTIVES

The final product of the facility characterization process is a clear statement or list of the protection objectives for the specific site. These objectives will be the metric used later to determine whether or not the protection system performs systematically to achieve the prescribed protection objectives. The protection objectives are derived by the security risk analysis team together with risk managers and are based on the information collected and modeled in facility characterization together with the specific security concerns.

Usually, the protection objectives for a building protection system include preserving the mission of the building. More specifically, protection objectives might include some subset of preventing:

- Disruption of normal work operations
- Compromise of the structural integrity of the building
- Compromise of the health and safety of occupants
- Disabling or misusing the utilities
- Disabling or misusing the HVAC
- Disabling or misusing the emergency systems

Additional protection objectives might also be associated with lower consequence events such as vandalism, petty theft, or embarrassment. The complete list of protection objectives must be specified before the remainder of the analysis can be completed, and the list must be kept consistent until at least the first iteration of the analysis is completed.

3.6   SUMMARY

A process has been presented for completing the facility characterization step for a security risk assessment. Facility characterization information is fundamental for completion of the risk assessment. The products that are required to complete further assessment include:

- List of undesired events
- Complete facility description
- List of critical assets to be protected to prevent undesired events
- List of protection objectives for the security system

The next chapter will discuss another important element of security risk assessment, which is the description of the adversarial threat to the facility.

3.7   REFERENCES

1. Biringer, Betty, "Risk Assessment Method for Electric Power Transmission," presented at Carnahan Conference on Security Technology, sponsored by IEEE, Albuquerque, NM, October 2004.
2. *Fault Tree Handbook*, U.S. Nuclear Regulatory Commission, NUREG-0492, January 1981.
3. Garcia, Mary Lynn, *The Design and Evaluation of Physical Protection Systems*, Butterworth-Heinemann, Burlington, MA, 2001.
4. Garcia, Mary Lynn, *Vulnerability Assessment of Physical Protection Systems*, Butterworth-Heinemann, Burlington, MA, 2006.
5. "The International Training Course for Nuclear Facilities and Materials," Sandia National Laboratories and the International Atomic Energy Agency, 30 April–19 May 2006, Albuquerque, NM.
6. Matalucci, Rudy and Strothman, John, "Security Risk Assessment Procedures: Countering Terrorism and Other Threats," Infrastructure Security Course sponsored by ASCE and Sandia National Laboratories, Las Vegas, NV, January 26–27, 2006.
7. Wyss, Gregory D. and Daniele, Sharon L., "Introduction to Fault Tree Analysis," ARRAMIS Training Course, Sandia National Laboratories, Albuquerque, NM.

3.8   EXERCISES

1. How are undesired events established for a specific site, organization, or industry?
2. List the types of information required to complete a facility description.
   a. What are some sources for the information?
   b. Define the list of participants and their roles in completing a facility description.
3. Why is it important to understand the workforce at a given facility before completing a security risk assessment?
4. What are the benefits for using a generic fault tree in facility characterization?
5. Discuss how to make the generic fault tree for buildings a site-specific fault tree.
   a. What are some applications for a site-specific fault tree?

    b. Discuss the relationship of critical assets to a site-specific fault tree.
6. List possible protection objectives for a typical commercial building, a military installation, and a government building.
7. Discuss the importance of having a complete facility characterization. Discuss possible ramifications of incomplete information.

Chapter 4

Threat Analysis

4.1 INTRODUCTION

A description of the adversarial threat is one of the most important components of any security analysis. The objectives of a threat analysis are to define the malevolent threat to a facility, corporation, region, or industry and to organize the collected data into a usable form that supports security risk management decisions. Threat analysis identifies and describes the types of adversaries (malevolent persons or groups) that may try to attack a particular facility. The most complete description possible of the malevolent adversarial threat spectrum is required. This description is used to estimate the relative likelihood that an adversary group will decide to attack a specific facility; the threat description is also key information used to estimate protection system effectiveness. Threat analysis provides this critical information.

Threat analysis is usually completed by a threat specialist with contacts in local, state, and national law enforcement organizations. Threat information is sensitive and must be protected accordingly because the threat level most often determines the level at which the facility will be protected. Threat analysis is a continuous process. After an initial analysis is completed, the threat description should be updated both periodically and as new adversarial threat information becomes available.

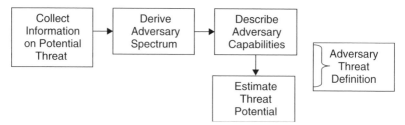

Figure 4.1   Threat Analysis Process.

Figure 4.1 depicts the process for conducting a threat analysis that is discussed in this chapter. Each step of this threat analysis process is developed in detail. This chapter provides a method and guidance for conducting a threat analysis by addressing how to:

- Collect information on the potential threat
- Derive an adversary spectrum for a given application
- Describe adversary capabilities
- Estimate the threat potential or likelihood of attack for specific adversary groups for a given asset
- Define the adversarial threat for a given entity

4.2   SOURCES OF THREAT INFORMATION

The threat analysis is usually completed by a threat specialist who has established ties and maintains contact with local, state, and federal law enforcement agencies, such as the Federal Bureau of Investigation (FBI). These and other sources of threat information are discussed below.

Threat information is sensitive. For the threat specialist to obtain threat information from any source, he or she must provide a credible need-to-know and instill confidence in the contacts that threat information will be properly protected and used with discretion. Information that identifies and describes the threat in a locale, region, and/or a specific industry or facility should be collected.

4.2.1 Local and State Sources

Local sources of threat information include the site's security manager, local law enforcement, and state or regional law enforcement. Site personnel who have been most successful in obtaining adversarial threat and local crime information are those who foster open and regular dialogue and a friendly relationship with local law enforcement officials. The advantage of inviting local law enforcement to the site for a tour of the facility and discussion – so that they know the facility and understand the security concerns – should never be underestimated. In many cases, law enforcement officers will share neighborhood and local crime information with site security managers.

Many states have organized threat and investigation working groups, formed primarily to address the terrorist threat. The concept of the Joint Terrorist Task Force (JTTF) has gained support in many states. In 1979, the New York City Police Department first used this concept of combining federal and local law enforcement capabilities in response to an overwhelming number of bank robberies. The concept of a joint effort proved to be valuable, so administrators eventually applied it to the counterterrorism program. The idea behind the establishment of the JTTF is a simple one: once established, the task force remains in place, becoming a close-knit, cohesive unit capable of addressing the complex problems inherent in terrorism investigations.

Following this JTTF concept, some industries with common interests formed alliances to meet and share threat information. Some of these groups established networks for reporting suspicious incidents to participating members. For example, in 1997 the Interagency Forum for Infrastructure Protection (IFIP) was formed to address a common task of meeting Presidential Decision Directive #63, which required all owners of critical infrastructure facilities to provide a plan for protecting their facilities against the international terrorist threat. The IFIP includes owners of federal dams and hydroelectric power producers. Members of IFIP include the Bonneville Power Administration (BPA), SNL, the U.S. Bureau of Reclamation (USBR), the U.S. Army Corp of Engineers (USACE), the Tennessee Valley Authority (TVA), the Western Area Power Administration (WAPA), the Federal Energy Regulatory Commission (FERC), and the FBI. This group developed security risk assessment methodologies for federal dams and high-voltage electric power transmission, and they continue to share lessons learned and threat information. The USACE has established a timely reporting and threat information-sharing system for unusual security incidents.

4.2.2   National Sources

Before September 11, 2001, sources for information on malevolent threats were limited to literature and Internet searches, crime studies, information sharing within professional organizations, and scant intelligence information. These sources continue to provide information on adversarial threats and crimes, but in the post-9/11 United States, the most timely and complete threat information is provided by the FBI and the Department of Homeland Security (DHS).

DHS provides information on the current threat level in the country through the Homeland Security Advisory system, which includes *Homeland Security Advisories* and *Homeland Security Information Bulletins*. The office of Science-based Threat Analysis and Countermeasures (STAC) in DHS has a mission objective to "improve our understanding of current and future threats" and a goal of characterizing and communicating current and future threats. One of the STAC Program Office's five core programs is Knowledge Discovery and Dissemination. This program implements a state-of-the-art information analysis architecture to acquire and maintain terrorist threat data and to provide real-time analysis and information processing for policy makers, intelligence analysts, law enforcement officials, human and animal health-care communities, and other decision makers.

The FBI operates the National Threat Center. The National Threat Center administers Counterterrorism Watch and other units related to threat management, such as:

- The Public Access Center Unit, which receives threat information from the public and forwards it to the appropriate unit
- The Terrorist Watch and Warning Unit, which produces finished intelligence products for the law enforcement community (such as *Intelligence Bulletins* and *Special Event Threat Assessments*)
- The Threat Monitoring Unit, which collects threat information, looks for patterns and connections, and provides "raw" threat-related intelligence reports

In addition, the FBI maintains the *National Security Threat List* (NSTL), which includes an *Issues Threat List* and a *Country Threat List*. The *Issues Threat List* is a list of eight categories of activity that represent a national security concern. These categories include terrorism, espionage, proliferation, economic espionage, targeting of the national information infrastructure, targeting the U.S. government, perception management, and foreign intelligence activities. The *Country Threat List* is a classified list of foreign powers that pose a strategic intelligence threat to U.S. security interests.

4.3   ADVERSARY SPECTRUM

The challenge is to organize all this collected adversarial threat data into a meaningful description that can be used in security analysis and made available to security risk decision makers. Because threat analysis is the process that answers two questions – Who is the threat? What is the level of severity of the threat? – the first step is to determine the types of adversaries in the threat spectrum.

In general terms, adversaries can be categorized as outsiders, insiders, or a collusion group. An outsider is a person who does not have official business with the facility and has not been granted routine access to the program, facility, or site. An insider is a person with authorized access to the facility or vital information about the facility; this person may be an employee, contractor, vendor, or visitor. The collusion group uses a secret agreement or cooperation between insiders and outsiders for illegal or malevolent purposes.

Outsiders can further be categorized as international terrorists, domestic terrorists, criminals, extremists, vandals, foreign intelligence personnel, and psychotics.

Terrorists – Terrorist groups have ideological, political, or issue-oriented motivations. They commonly work in small, well-organized groups or cells and can be very sophisticated in their skill levels. Most terrorist groups have technical training, are skilled with tools and weapons, and plan efficiently. International terrorists include groups like Al Qaeda or Hamas; domestic terrorists might include ecological terrorists, white supremacist groups, and the more violent environmental or animal activist groups.

Criminals – Criminals are persons who commit criminal acts for profit or economic gain. White-collar criminals are individuals who seek classified and/or proprietary information for the purpose of gaining economic advantage for the individual or the individual's employer.

Extremists – Extremists work in small, well-organized groups. They are politically or issue-oriented, acting out of frustration, discontent, or anger against other social or political groups. Their goals range from publicity to damage and destruction; most extremists are mere protesters, those that are more violent and destructive, such as ecological terrorists, are considered to be in the terrorist category.

Vandals – Vandals are unsophisticated and superficially destructive in nature. They do not intend to injure people or cause extensive damage to targets. Their targets are usually targets of opportunity – whatever happens to be located in the vicinity of where they are. Computer hackers could be considered a type of vandal.

Foreign intelligence personnel – The foreign intelligence officer is an individual who uses human intelligence methods and engages in clandestine intelligence gathering on behalf of a foreign intelligence service. His or her primary function is to collect information and/or recruit insider assistance.

Psychotics – The psychotic is a person suffering from a mental disorder of sufficient magnitude to experience periodic or prolonged loss of contact with reality. Psychotics can fit in the outsider or insider parts of the threat spectrum.

An *insider* can be defined as anyone with knowledge of operation or security systems and who has unescorted access to facilities or security interests. The insider threat can range from a person who is passive (only provides information) or active nonviolent (facilitates entrance and exit, disables alarms and communications, and the like) to the active violent (actively participates in a violent attack). Although having more than one insider is possible, emphasis is placed on addressing the single insider, the most probable insider threat.

Addressing the insider threat can be difficult because of the social sensitivities (of employees feeling like they are being watched or suspected) and the challenge of protecting against an insider whose very job assignment grants him or her authorized access to critical assets and/or the equipment that protects and monitors the critical assets. A more acceptable analytic method for addressing insider threats is to identify all of the types of employment positions at the facility and then consider the access and authority advantages afforded by each position. The analysis can become more objective when the question considered is *IF* there were an insider in this employment position, how effective is the protection system? The focus becomes how often does someone in this job position have lone access to the critical asset, lone access to the protection system or monitoring system of the asset, or opportunity to collude with an outsider. At some point, management is faced with the likelihood that one of their employees might decide to become an insider adversary. For managers of critical operations or high-consequence facilities, considering the threat potential of an insider threat must be addressed. For many facilities, certain employment positions afford individuals 'the keys to the kingdom.' This unlimited access must be considered when analyzing the insider threat.

The *collusion threat* usually takes the form of a single insider supporting a group of outsiders. An active insider participates fully in the attack; a passive insider provides information about the facility operations and safeguards to a group of outsiders.

4.4   ADVERSARY CAPABILITY

For each adversary group that is identified, the most complete description possible is developed for use in designing and/or evaluating the effectiveness of a protection system. The type of information sought in the data collection task for each adversary group includes:

- Motivation
- Tactics
- Intelligence-gathering means
- Targets of interest
- Expected number in group
- Equipment
- Transportation
- Weapons

- Explosives
- Technical skills/knowledge
- Financial resources
- Potential for collusion with an insider

Motivation – This is generally based on goals and objectives. Threat motivations are described as economic (desire for financial gain), ideological (linked to political or philosophical system), or personal (related to special situation of specific individual – hostility, grievance/revenge, psychotic, economic gain). Motivation is related to the desired level of consequences: make a statement, cause mass casualties, create terror, or cause economic crisis.

Tactics – This describes the type of tactics the group has used or might be expected to use. Tactics include a suicide bomb, a standoff attack, an intrusion, sabotage, destruction, and/or cyber-attack.

Intelligence-gathering means – This addresses how the group gets information, namely through documentation, human intelligence, signal intelligence, photographic intelligence, and/or physical or cyber-surveillance.

Targets of interest – This demonstrates past or current interests: financial institutions, critical infrastructure elements, recreational sites, religious facilities, or a specific industry (chemical facility, animal research institute, or the like).

Expected number in group – The total number of adversaries expected in an attack and whether or not the group is broken up into smaller groups or cell.

Equipment – This includes the types of equipment that the group both has access to and has the skill to operate: heavy construction equipment, hand tools, power tools.

Transportation – This describes the type of transportation that the group might use in an attack: aircraft (jet liner, helicopter, ultra-light), boat, motor vehicle (truck, van, sedan).

Weapons – This lists the types of weapons that the group has access to and is trained to use: handguns, rifles, shotguns, automatic weapons.

Explosives – This describes the types of explosives that the group might be expected to use: bulk, shaped charges, improvised explosives.

Technical skills/knowledge – This includes technical training and experience: engineers, scientists, special combat forces, explosives handlers.

Financial resources – This describes the source of money to conduct attacks: self-financed, state-sponsored, organized crime.

Potential for collusion with insider – This predicts the potential for collusion with an insider or an interest in colluding with an insider.

After all threat information for each adversary group has been collected, a table, like the example in Table 4.1, can be constructed to organize the information. The table is valuable for risk managers and security analysts who design and/or evaluate security system effectiveness. In addition, the summarized threat information can be used to assess the threat potential for attack for a given adversary group and facility.

4.5   THREAT POTENTIAL FOR ATTACK

For security risk analysis, unlike safety risk analysis, estimating the likelihood of the initiating event is difficult at best. Ideally, security and safety risk analyses could share methodologies for estimating risk parameters. Unfortunately the differences between these two analyses are too significant for a shared methodology. A primary difference can be found when estimating the likelihood of the initiating event. Safety studies have historical data and statistics that predict the likelihood of an abnormal event and the system's response to the event. For security studies, estimating the likelihood that an adversary group will attack a specific asset

Table 4.1 Hypothetical Adversarial Threat Summary

Type of Adversary	No. of Adversaries	Equipment	Vehicles	Weapons	Motivation	Tactics	Targets of Interest
International Terrorist (May include an insider colluding)	3–10	Unlimited hand/power tools, body armor, chem/bio, wireless comm.	Car, pickup, 4 × 4, truck, semi-trailer truck, boat, barge, aircraft	Handguns, automatics,[b] explosives,[c, d] chem/bio	Mass casualties, widespread fear, psychological impact, economic crisis	Catastrophic event, sniper, hostage, cyber	People, high-visibility national events, critical infrastructure, soft targets, national icons
Domestic Terrorist: Eco & Militia (May include an insider colluding)	3–5	Hand and power tools, body armor, chem/bio	All-terrain vehicle, car, pickup, 4 × 4, truck, boat, aircraft	Handguns, automatics,[b] explosives,[c, d] chem/bio, incendiary devices	Retaliate against the government, make a political statement, change business practice	Catastrophic event, hostage, sniper, arson, cyber	Specific government facilities, facilities with environmental issues, people

Table 4.1 (continued)

Type of Adversary	No. of Adversaries	Equipment	Vehicles	Weapons	Motivation	Tactics	Targets of Interest
Criminal	2–3	Hand tools	Car, pickup, 4 × 4, truck	Handguns, automatics,[b] knives	Financial gain. Steal property.	Property theft, cyber	High-economic-value assets, such as banks
Extremists	5–10	Signs, chains, locks, hand tools	Car, pickup, 4 × 4, truck, boat, bus	Incendiary devices, clubs	Make a political statement, protest	Protest, civil disobedience, assault, damage, destruction, cyber	Facilities with political or environmental significance.
Vandals/ Hackers	2–5	Spray paint, rocks, knifes	Car, pickup	Handguns, automatics[b]	Vandalism	Damage, destruction, hacking (cyber)	Conveniently located facilities.

Table 4.1 (continued)

Type of Adversary	No. of Adversaries	Equipment	Vehicles	Weapons	Motivation	Tactics	Targets of Interest
Foreign Intel. Officer	1	Human intelligence-gathering strategies	Car	Usually none	Gain access to classification/proprietary info, recruit insider	Cyber, surveill., approach insider employee	Information, products, insider employee
Psychotic	1	Hand tools	Car	Up to handguns	Random	Violence, destruction	Random
Insider[a]	1	On-site equipment	Car, pickup, 4 × 4, boat	Hand guns, etc.	Disgruntled	Destruction, violence, theft, cyber	Facility asset(s) and equipment, fellow employee(s).

[a] Insiders include special-interest groups such as employees (present and former), contractors, or vendors.

[b] Type of weaponry includes shotguns, rifles with long-range capabilities, automatic weapons, and large caliber weapons.

[c] Explosives (hand-carried) would be limited to what one person could carry, pipe bomb or backpack bomb.

[d] Explosives (vehicle-carried), car, pickup, truck or semi-trailer truck.

presents a challenge because of the human element. Humans plan, rehearse, learn, and modify in order to optimize the attack's effectiveness; the events are not random, and many of the required mathematical assumptions cannot be met. Human behavior is difficult to predict and providing a quantified prediction of human behavior is an even more difficult task.

Historical adversary attack data, even if it were available, would not necessarily predict whether an adversary group would attack a particular facility/asset. Adversaries, especially those motivated by ideology, have very subjective reasons for attacking a particular target and they gather data, conduct surveillance, and rehearse until they are confident of their success. Even though likelihood of attack is difficult to quantify, adversaries must go through some logical process to determine the target. The logical process could vary because of variations in motivation, capability, and tactics among adversary groups.

The process for estimating the threat potential follows a complete threat definition. Estimating threat potential is an attempt to estimate the likelihood that an adversary group would decide to attack a particular facility or entity and will later be referred to as likelihood of attack. For the insider threat, threat potential is the likelihood that an insider is an adversary.

4.5.1   Outsider Threat

A qualitative relative threat potential parameter is used to replace the likelihood of adversary attack for a specific facility. The method to estimate the threat potential follows a complete threat definition and the parameter is estimated for a specific facility and adversary group. The parameter is based on characteristics of this adversary group, such as capability, historical actions, and current intent, as well as target attractiveness of the facility/asset to the adversary group.

The assessment method for estimating threat potential produces a qualitative threat potential parameter for a given adversary

Table 4.2 Factors for Estimating Threat Potential

Adversary Capability	Adversary History/ Intent	Relative Attractiveness of Asset to Adversary
• Access to region • Material resources • Technical skills • Planning/organizational skills • Financial resources	• Historic interest • Historic attacks • Current interest in site • Current surveillance • Documented threats	• Desired level of consequence • Ideology • Ease of attack

group and undesired event for a facility/asset. The factors used for estimating threat potential are divided into three sections and are illustrated in Table 4.2. The first factor is a Yes/No question (Is the adversary capable of the attack?) that assesses the threat potential as very *Low* or *High*. The second and third factors are considered after a Yes answer to the capability question and are evaluated individually by a scoring process and then combined (summed) to provide a relative score for adversarial threat potential.

The first factor addresses adversary capability. Capability establishes whether the likelihood of attack is very low or greater. If adversaries are *not* judged to be capable of causing an undesired event at the facility and achieving the level of consequences, the threat potential is judged to be very low. The other two factors further refine the likelihood of attack. The second section addresses the history/intent for adversary groups that are judged to be capable of causing the undesired event. Finally, the third section considers the relative attractiveness of the target to the specific adversary group.

Capability determines whether the adversary is assessed to have or has demonstrated the capability to conduct an attack on the facility. Adversaries are usually described in terms of the degree of capability they have in several categories. A non-zero degree of capability in every category is necessary; whether an adversary has these aspects of capability can usually be discovered. An analyst can deduce limits on the severity of attack that an adversary can mount by the degree of capability. Based on the threat data gathered, the first questions to ask are whether the adversary group is located near or able to gain access to the region and whether they are expected to have the material resources, technical skills, planning/organization skills, and financial resources to attack the facility. If the answer is *No*, then threat potential is judged to be Very Low. If the answer is *Yes*, the analysis continues. Table 4.3 summarizes this first step in the assessment method.

If the specified adversary group is judged to be capable of a successful attack on the facility, the threat potential is estimated by a relative score based on *History / Intent* and the *Relative*

Table 4.3 Assessing Adversary Capability

Capability: Is the adversary group capable of conducting a successful attack on this facility? To answer the question, consider the threat description. Is the adversary group:	If YES, continue	If NO, Threat Potential Is Very Low Stop
Located near or able to gain access to the region?		
Expected to have the material resources to attack this facility?		
Expected to have the technical skills to attack this facility?		
Expected to have the planning/ organizational skills to attack this facility?		
Expected to have the financial resources to attack this facility?		

Attractiveness of the Facility / Asset to the adversary group. Scores are associated with the answer that most accurately describes specific items of interest. If it is judged that some items have a greater impact on threat potential, they can be assigned a relatively higher score.

History/Intent – The cumulative score for History/Intent is based on historic and current interests of the adversary group. History is captured by past interests (evidence that this adversary group has shown interest in this type of facility or this specific facility) and attacks (evidence that this adversary group has conducted similar attacks in the past at this facility or this type of facility). Current interest is described by current interest in the facility (information suggests interest in the facility), current surveillance (existence of intelligence information regarding this site or similar facility), and documented threats (existence of documented threats to the facility). Table 4.4 summarizes this second step of the method.

Relative Attractiveness of the Facility/Asset – The scope for Relative Attractiveness of the Facility/Asset is based on attributes of the facility relative to the interests of the adversary group. Relative Attractiveness of the Facility/Asset is captured by consequence (whether or not the estimated level of consequence for the attack is consistent with goals of the adversary group), ideology (whether or not attacking this facility is consistent with the ideology/motivations of this adversary group), and relative ease of attack (perception of how easy it is to defeat the protection system and/or how easily the undesired event can be accomplished). Table 4.5 summarizes this step of the assessment method.

Threat potential is then estimated by first summing all scores for History/Intent and Relative Attractiveness of the Facility/Asset. The total scores can be partitioned into bands to estimate a threat potential factor per adversary group and facility/asset. For

Table 4.4 History/Intent Assessment

History / Intent				Score
Historic interest	If there is documented evidence that historically this adversary group has shown interest in this type of facility or this specific facility, Score = 5.	If there is speculation, but no evidence that this adversary group has shown interest in this type of facility or this specific facility, Score = 3.	If there is no evidence that this adversary group has ever shown interest in this type of facility or this specific facility, Score = 1.	
Historic attacks	If there is documented evidence that this adversary group has conducted similar attacks in the past at this facility or this type of facility, Score = 5.	If there is speculation, but no evidence that this adversary group conducted similar attacks in the past at this facility or this type of facility, Score = 3.	If there is no evidence that this adversary group has conducted similar attacks in the past at this facility or this type of facility, Score = 1.	
Current interest in facility	If current information suggests interest in the facility, Score = 10.	. . .	If there is no current information that suggests interest in facility, Score = 2.	

Table 4.4 (*continued*)

History / Intent				Score
Current surveillance	If current intelligence documents surveillance at specific facility, Score = 10.	If current intelligence documents surveillance at other similar facilities in the U.S. or other types of facilities in the region, Score = 6.	If current intelligence does not involve specific facility, similar facilities in the U.S. or other types of facilities in the region, Score = 2.	
Documented threats	If this facility has received documented threats of attack from this adversary group, Score = 10.	If this facility has received documented threats of attack but not from this particular adversary group but similar groups, Score = 6.	If this facility has not received documented threats from this adversary group or other adversary groups, Score = 2.	

example, if total scores for a given adversary group and facility are estimated to be:

- Greater than V and less than or equal to W, then threat potential is *Low*
- Greater than W and less than or equal to X, then threat potential is *Medium*

- Greater than X and less than or equal to Y, then threat potential is *High*
- Greater than Y and less than or equal to Z, then threat potential is *Very High*

The values of the bounds should be predetermined for the given industry or type of facility and based on the expert judgment of the threat analyst. These values must be consistent throughout the analysis. The estimated threat potential parameter, though relative in nature, can provide valuable information to security

Table 4.5   Relative Attractiveness of the Facility/Asset Assessment

Relative Attractiveness As Target				*Score*
Consequence	If level of estimated consequence for attack is consistent with goals of this adversary group, Score = 10.	If level of consequence caused by attack is not definitely consistent with goals of adversary group, but possibility exists, Score = 6.	If level of consequence caused by attack is not at all consistent with goals of this adversary group, Score = 2.	
Ideology	If attacking this facility is consistent with ideology/ motivations of this adversary group, Score = 10.	If attacking this facility is not consistent with ideology/ motivations of this adversary group but possibility exists, Score = 6.	If attacking this facility is not at all consistent with ideology/ motivations of this adversary group, Score = 2.	

Table 4.5 (*continued*)

Relative Attractiveness As Target				*Score*
Ease of attack	If perception exists that PPS is relatively easy to defeat or doesn't exist and/or the undesired event is easily accomplished at this facility, Score = 5.	If perception exists that the PPS at the facility provides moderate protection and/or there is moderate difficulty in accomplishing the undesired event at this facility, Score = 3.	If perception exists that the facility has a robust, effective protection system and/or the undesired event is extremely difficult to accomplish at this facility, Score = 1.	

risk managers. The parameter can be used in conjunction with consequence levels to prioritize facilities/assets and/or to estimate relative security risk level. The net result is logical guidance to optimize the allocation of limited security resources.

4.5.2 Insider Threat

Estimating the threat potential for the insider threat is probably the most daunting task of security risk managers. The task is socially, politically, and legally sensitive, and it is technically challenging to protect against the trusted insider. Predicting what would make an employee decide to become an adversary is difficult, at best. For high-risk employment positions, those which require significant access to sensitive or proprietary information, materials, products, cyber-systems, or the protection systems for these items, consideration of the threat potential is of paramount importance.

The most effective protection for an insider must occur before or while the employee is making the decision to become an adversary. Once the decision is made or the employee recruited, the insider may go undetected. History has shown that most spies are never detected or caught in the act; rather they are reported by other spies. A physical protection system (PPS) can make it difficult for the insider to do the wrong thing, but it cannot act alone to completely mitigate the insider threat. Protection against the insider threat requires an integrated protection system with personnel screening, physical protection, cyber-protection, and operations security. Appendix D, "Insider Threat" discusses protection concepts for the insider threat.

Personnel screening is the one protection function that can occur during the insider decision-making or pre-recruitment (by a malevolent group) phase. Some type of personnel screening should be conducted for positions judged to be high risk. Pre-employment screening and continuous updates should be conducted to ensure the protection function's effectiveness. Screening methods must be reviewed by the labor counsel and must respect the personal privacy of employees. Turner and Gelles suggest a list of characteristics that make individuals attractive targets for recruitment by outside malevolent groups, such as foreign intelligence services or terrorist groups. The characteristics that they suggest for an effective pre-employment screening include:

- Alcohol or other substance abuse
- Financial issues
- Criminal behavior (including juvenile)
- Workplace performance or behavior (previous)
- Compulsive or excessive gambling
- Repeated policy violations (rules do not apply to them)

4.6   SUMMARY

Threat analysis attempts to answer the questions: Who is the threat? How strong are they? What is the likelihood that they will decide to attack my facility? The detailed answers to these questions are used to design and/or evaluate the effectiveness of a protection system. The level of required security protection is dependent on the level of the potential threat. With limited financial resources for security systems, security risk decision makers must have the best threat information available and a logical method for prioritizing the adversarial threats to their facilities. This chapter has described the importance of a timely (current) and complete threat definition. A process for conducting a threat analysis was provided by offering guidance on where to obtain threat information, how to organize the information to make it usable for system effectiveness analysis, and how risk managers can use it to make security decisions.

4.7   REFERENCES

1. Biringer, Betty, "Estimating Threat Potential for Security Risk Analysis," presented at American Nuclear Society Conference, San Francisco, CA, September 2005.
2. Department of Homeland Security, "Threats and Protection: Synthesis and Dissemination of Information," http://www.-dhs.gov/dhpublic/theme_home6.jsp.
3. *FBI's Counterterrorism Report, Since September 2001*, A Report to the National Commission on Terrorist Attacks upon the United States.
4. Garcia, Mary Lynn, *The Design and Evaluation of Physical Protection Systems*, Butterworth-Heinemann, Burlington, MA, 2001.
5. *International Training Course for Nuclear Facilities and Materials – Volume I. Determining Physical Protection System Objectives*, Sandia National Laboratories and the International Atomic Energy Agency, 2004.

6. National Counterterrorism Center (NCTC) Knowledge-Based Directory, http:/www.tkb.org/Home.jsp.

7. *National Security Threat List (NSTL),* Federal Bureau of Investigation, http://www.dss.mil/training/csg/security/T1threat /Nstl.htm.

8. Parker, Dr. Gerald, *Mission and Goals of the Office of Science-Based Threat Analysis and Counterterrorism,* Department of Homeland Security, November 2004.

9. Paulus, William K., Briefing Package, *Non-nuclear Risk Assessment Methodology: Threat Assessment,* Sandia National Laboratories, February 2002.

10. Turner, James T., PhD and Gelles, Michael G., PsyD, *Threat Assessment: A Risk Management Approach,* Haworth Press, Inc., Binghamton, NY, 2003.

4.8   EXERCISES

1. List two or more uses of *threat analysis* in security risk analysis. Discuss possible ramifications if the *analysis* is incomplete.

2. When should a threat analysis be completed for a given facility or industry? For what period of time is the threat description valid?

3. Threat information can be collected from various dependable sources.
   a. List possible local sources
   b. List state or national sources
   c. List others (Where else might you look for threat information?)

4. Assume that you are the threat specialist and have collected data from every source that you can find for your facility.
   a. Discuss how you would decide which adversary groups you would include in the adversary spectrum to present to management.
   b. What would you do or what advice would you provide to management if you could find no threat information that you considered relevant for your facility? Specifically, what would you advise in the absence of threat data. Why?

5. A list of potential adversary capabilities was provided in this chapter. List any others that you think might be important. Based

on your judgment, rank order the top five capabilities and discuss why you ranked them in that order.

6. What are the challenges to estimating the likelihood of adversary attack? Can this estimate be quantified? Why or why not?

7. The insider threat poses a great challenge to a protection system. Why is protection for the insider threat so important?

8. The design threat is the term used for the site-specific threat spectrum that management decides to employ to design the security system for that facility. The level of protection will be established by this specific spectrum. Discuss the differences in level of protection for a security system that is designed to meet a full threat spectrum (terrorists, criminals, extremists, vandals, insiders) versus a security system that is designed to meet criminal and vandal adversary groups.

Chapter 5

Consequence Analysis

5.1   INTRODUCTION

In Chapter 3, "Facility Characterization," the identification of the undesired events and the critical assets that must be protected to prevent the undesired events was discussed. The purpose of consequence analysis is to estimate consequence values for each undesired event for a given facility. The consequence values are later used to estimate relative security risk. The basic process to complete consequence analysis is:

- Determination of a reference table of consequences
- Estimation of consequence values for undesired events

5.2   REFERENCE TABLE OF CONSEQUENCES

In order to establish a standard for discussing the consequences associated with a specific undesired event, a *reference table of consequences* is developed. The table is populated by the security analysis team and technical experts for final approval by management. The contents of the table should remain constant throughout the analysis. If a comparison or relative ranking of facilities is to be completed, the same reference consequence table must be used to estimate consequence values at each facility being compared.

The first components of the reference table of consequences are the criteria for describing consequence. Criteria may be site-, organization-, or industry-specific. Measurable, rather than subjective, criteria should be used. Common criteria or units of consequence include:

- Deaths
- Economic loss (to owner)
- Economic loss (to customer)
- Loss of operations or production
- Loss of public confidence
- Loss of asset(s)
- Downtime
- Geographic impact
- Population at risk

The next step is to establish definitions for qualitative values of each consequence criterion. As many levels of consequence severity can be used as can be discretely defined. For example, assume that loss of production is a selected criterion. Consequence value definitions for this criterion might look like:

Undesired Event	Consequence Criteria	High	Medium	Low
Loss of Production Capability	Future contracts	Severe	Moderate	Low
	Duration	>2 Years	1–2 Years	<1 Year
	Economic Loss	>$5 million	$1–$5 million	< $1 million

Table 5.1 DoD Military Standard 882D

Consequence Category	Consequence Level
Could result in death, permanent total disability, loss exceeding $1M, or irreversible severe environmental damage that violates law or regulation.	Catastrophic
Could result in permanent partial disability, injuries, or occupational illness that may result in hospitalization of at least three personnel, loss exceeding $200 K but less than $1M, or reversible environmental damage causing a violation of law or regulation.	Critical
Could result in injury or occupational illness resulting in one or more lost workday(s), loss exceeding $10 K but less than $200 K, or mitigatible environmental damage without violation of law or regulation, where restoration activities can be accomplished.	Marginal
Could result in injury or illness not resulting in a lost workday, loss exceeding $2 K but less than $10 K, or minimal environmental damage not violating law or regulation.	Negligible

Table 5.1 repeats Table 1.1 from "Chapter 1, Security Risk Assessment," which provides the Department of Defense Military Standard 882D for consequences that can be used as a start for a reference table of consequences. Table 5.2 shows a hypothetical example of a reference table of consequences.

5.3 CONSEQUENCE VALUES FOR UNDESIRED EVENTS

Once the reference table for consequences has been determined, each undesired event that was identified in the facility characterization step is analyzed for the consequence severity that

Table 5.2 Hypothetical Example of Reference Table of Consequences

Measure of Consequence	High	Medium	Low
Economic loss (property loss + revenue)	> $5M	$1–5M	< $1M
Economic loss (users)	> $5M	$1–5M	< $1M
Deaths	> 3	1–3	0
Geographic Impact	National	Regional	Local
Public confidence			
Intra-regional	Outage > 1 week	Outage > 1 day but < 1 week	Outage < 1 day

would result if the undesired event did in fact occur. Criteria and definitions in the reference table of consequences are used to estimate the consequence severity. If more than one criterion is used to describe consequence, each criterion is evaluated and then the highest consequence value is selected to estimate the consequences of the undesired event.

Table 5.3 shows an example of estimating consequences for undesired events. Consequences are estimated by the analysis team with input from technical experts. The safety analysis team may be able to support this consequence estimating effort. However, a caution is that if credit is attributed to safety features to reduce the consequences of an event, there must be confidence that the safety features would survive a malevolent attack. In other words, the adversary may plan to defeat the safety features as part of the malevolent attack on the facility in order to maximize consequences.

Table 5.3 Consequence Value Estimation Example

Undesired Event	Measure of Consequence		Consequence Severity	
	Type	Value	By Type H/M/L	By Event H/M/L
Disruption of Operations (sabotage of vital equipment by cyber-attack)	Economic loss (property loss + revenue)	$3M	M	
	Economic loss (users)	0	L	
	Deaths	0	L	
	Geographic impact	Local	L	
	Public confidence	6 months	H	
			Enter highest consequence	H
Theft of Valuable Asset(s) (precious metals)	Economic loss (property loss + revenue)	$1M	M	
	Economic loss (users)	0	L	
	Deaths	0	L	
	Geographic impact	Local	L	
	Public confidence	1 day	L	
			Enter highest consequence	M

(continued overleaf)

Table 5.3 (*continued*)

| Undesired Event | Measure of Consequence | | Consequence Severity | |
	Type	Value	By Type H/M/L	By Event H/M/L
Crimes Against People (hostage situation)	Economic loss (property loss + revenue)	0	L	
	Economic loss (users)	0	L	
	Deaths	0–1	L	
	Geographic impact	Local	L	
	Public confidence	None	L	
			Enter highest consequence	L
Destruction of Building (vehicle bomb)	Economic loss (property loss + revenue)	$7M	H	
	Economic loss (users)	0	L	
	Deaths	10–20	H	
	Geographic impact	Local	L	
	Public confidence	6 months	H	
			Enter highest consequence	H

Other important issues in estimating consequences are keeping the criteria and definitions constant during the analysis and documenting of any assumptions made in estimating consequence severity. Examples of assumptions might be whether or not hardware replacement costs are included in the estimates or if the "domino effect" is included if other facilities could be affected by the attack or if contingency operations during recovery are included or not. Consequences may be scenario-dependent. To be security conservative, extreme (most severe) conditions should be used as "bounding" measures.

5.4 SUMMARY

The products of a Consequence Analysis are a site-specific reference table of consequences and relative consequence values for undesired events. Consequence values will be used together with threat and protection system effectiveness values to estimate relative security risk for the list of undesired events.

5.5 REFERENCES

1. Biringer, Betty, "Risk Assessment Method for Electric Power Transmission," presented at Carnahan Conference on Security Technology, sponsored by IEEE, Albuquerque, NM, October 2004.
2. MIL-STD-882D, "Department of Defense Standard Practice for System Safety," February 10, 2000.

5.6 EXERCISES

1. Why should the reference table of consequences be site/organization /industry specific?
2. Assume you have a multistory office building that houses six different companies, including one government organization. List the possible consequence criteria.
3. What are the limitations of having more than three severity levels for consequence criteria (normally high, medium, and low)?
4. Discuss the makeup and expertise of the team required to complete consequence analysis.
5. What are some important assumptions that might affect the consequence analysis outcome?

Chapter 6

Asset Prioritization

6.1   INTRODUCTION

At this point in the security risk assessment process, the threat parameter and the consequence parameter have been estimated. Corporations or organizations that own many facilities may be faced with the dilemma of limited resources – time and or money – to complete risk assessments for all of their assets. The threat parameter together with the consequence parameter can be used to prioritize or order assets in a given facility or building in terms of which might be at the highest risk so that owners can address them first. Many corporations do not require a prioritization step, but for other corporations, prioritization is especially valuable. This prioritization scheme was used as a higher-level "screening" process to prioritize facilities. For example, in the year 2000, federal dam owners in the United States began to do security risk assessments for their dams. After September 11, 2001, there was an urgency to complete the assessments as quickly as possible and to address the most critical dams first. With more than 75,000 federal dams in this country, the prioritization scheme presented in this chapter helped owners decide which dams to analyze first and further helped owners schedule the assessments for the remainder of their dams in the years following.

6.2   PRIORITIZATION MATRIX

The prioritization matrix is constructed by plotting the ordered pairs of threat potential (likelihood of attack) vs. consequence (of successful attack) for assets for the most critical site-specific adversary group. The matrix can be used to prioritize either a number of different undesired events for a given facility or to prioritize different facilities.

Figure 6.1 provides an example prioritization matrix for undesired events at an example facility. Note that the shaded area of the matrix highlights those undesired events/assets with medium or higher values for both threat potential and consequence. Figure 6.2 provides an example of a prioritization for buildings owned by a corporation.

The interpretation of the prioritization matrix for planning purposes is an exercise for the analysis team and the risk managers. Obviously, the assets shown in the upper-right corner of the matrix pose the highest risk because they have the highest likelihood of being attacked and the highest consequences if the attack is successful. The matrix should be used to schedule security risk assessments for the given time and money resource constraints. The order in which the assessments are done must

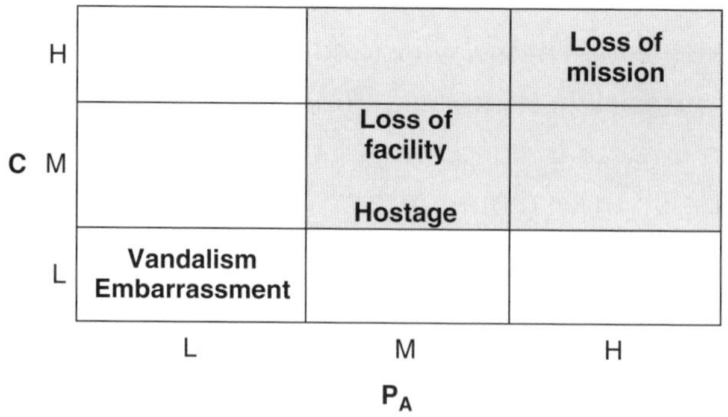

Figure 6.1   Prioritization Matrix Example for Undesired Events/Assets at a Facility.

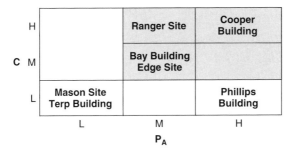

Figure 6.2 Prioritization Matrix Example for Sites.

address site-specific conditions. The prioritization matrix should *not* be used to eliminate some assets from undergoing a security risk assessment.

6.3 SUMMARY

The security risk equation threat parameter together with the consequence parameter for a given asset can be used to prioritize assets. Assets with *Medium* or higher likelihood of attack and consequence value can be identified. Security risk managers can use this information to order assets for a full security risk assessment. The prioritization step is optional. Some owners only have a few assets and can easily complete security risk assessments for all of them, while others have numerous facilities and cannot complete risk assessments at all of them in a timely manner. Prioritization helps these owners decide which assets to address first.

6.4 REFERENCES

1. Biringer, Betty, "Risk Assessment Method for Electric Power Transmission," presented at Carnahan Conference on Security Technology, sponsored by IEEE, Albuquerque, NM, October 2004.
2. "Understanding Risk in a Changing World," Short Course taught at Society of Women Engineers National Conference, October 16, 2004, Milwaukee, WI.

6.5   EXERCISES

Assume a corporation has completed threat and consequence analysis for malevolent attacks at its seven buildings. The results are tabulated below.

Building	Likelihood of Attack	Consequence
A	H	M
B	M	L
C	M	M
D	H	H
E	H	H
F	L	M
G	M	L

1. Construct the prioritization matrix:

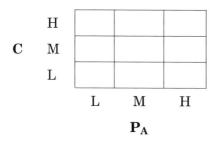

2. Suggest the order for security risk assessments to be completed for the buildings if the owner has three years to complete the assessments and has the resources to do two or three per year. Provide reasons for your ordering and note any assumptions that you made.

3. What are some other events, factors, or conditions that might alter the assessment schedule?

4. How does *prioritization of assets* compare to the *screening analysis* discussed in Chapter 2? When would you use screening? When would you use asset prioritization?

Chapter 7

System Effectiveness

7.1   INTRODUCTION

The purpose of *system effectiveness assessment* is twofold:

1. Estimation of protection system effectiveness
2. Identification of site-specific vulnerabilities

System effectiveness assessment begins with a review of the information derived in facility characterization and threat analysis discussed in previous chapters. The specific required information includes:

- Site-specific fault tree
- PPS description   Physical Protection System
- Cyber-protection system description
- System protection objectives
- Threat description

In estimating protection system effectiveness, a systematic approach is used to answer the basic question, "To what level does the protection system meet the required protection objectives for the given adversarial threat?" The desired situation is when the answer to the question is *High* or *Very High*. When the answer is *Medium* or lower, the next question is "What makes the system protection effectiveness *Medium* or *Low*?" The weaknesses or gaps in protection that result in the lower effectiveness level represent site-specific vulnerabilities.

7.2 PROTECTION SYSTEM EFFECTIVENESS

A systematic approach to estimate protection system effectiveness is used. The steps include:

- Identification of adversary strategies
- Assessment of PPS effectiveness *physical Protection system*
- Assessment of cyber-protection system effectiveness

7.2.1 Adversary Strategies

Adversary strategies to accomplish undesired events at a site are outlined in the site-specific fault tree, which is why it is suggested that once the generic tree is made site-specific, the tree should be adequately protected. In an attempt to "bound" the problem, it is advantageous to use the most-vulnerable strategy for the analysis. The most-vulnerable strategy is defined as the strategy that provides the greatest advantage to the adversary to accomplish the undesired event. Selection of the most-vulnerable strategy is made by using expert (team) opinion based on knowledge of the site, the operations, the information systems (process control and SCADA), and the existing physical and cyber-protection system features. Both physical and cyber-strategies should be considered for each undesired event.

The most-vulnerable strategy to accomplish an undesired event must be identified first. Several factors must be considered in judging which strategy might be (relatively) the most-vulnerable:

- Protection system weaknesses noted during facility characterization
  - Least-protected physical system features (detection, delay, response)

SCADA = Supervision Control and Data Acquisition

- - Least-protected cyber-protection features (authentication, authorization, audit)
- Easiest system features to defeat
- High consequence results
- Facility operating states that the adversary could use to an advantage
  - Emergency conditions
  - Above normal operating level
  - No personnel on-site
- Inclement weather
- Loss of power source(s), communications, or other support infrastructure

If the decision of which strategy among others is "most-vulnerable" is too difficult to make, more strategies should be addressed for the undesired event. As many strategies as needed should be developed to provide confidence in the judgment. If both physical and cyber-attacks are possible for a given undesired event, the analysis should be completed and reported for both types of attacks.

For demonstration purposes, assume a priority undesired event is the loss of mission. From the fault tree, disruption of building mission can be accomplished by disrupting normal work operations, compromising the structural integrity of the building, compromising the health and safety of occupants, the disablement or misuse of utilities, the disablement or misuse of the HVAC, or disablement or misuse of emergency systems. For demonstration purposes, assume that the information gathered in facility characterization leads the analysis team to judge that disrupting work operations by destroying or manipulating the information system would be the adversary strategy of choice. Specifically, assume that based on relative level of consequence, known protection system weaknesses, and ease of adversary attack, the team decides that the *most-vulnerable strategy to cause loss of mission is to disrupt normal work operations by an attack on the information system.*

7.2.1.1   Critical Assets for Strategies

Once the most-vulnerable strategy(ies) for an undesired event is defined, the next step is to document the critical asset(s) associated with the strategy. Protection of these critical assets will be the focus of the analysis. Both physical and cyber-attack options should be considered for each critical asset. In the above example, assume that the critical assets would be *the computer equipment in the control room (for a physical attack) and the electronic communications to the process control equipment for the information system (for a cyber-attack)*.

In the next section, construction of an adversary sequence diagram (ASD) for each critical asset (associated with the most-vulnerable strategy) will be discussed.

7.2.2   Physical Protection System Effectiveness

In this section, a systematic method developed and used extensively by SNL for estimating PPS effectiveness will be demonstrated. For more than 20 years, the basic method has been applied to numerous types of PPSs. Several tasks will be described to assess PPS effectiveness:

- Development of an ASD for each critical asset
- Association of detection and delay values for each element of the ASD
- Selection of the most-vulnerable scenario
- Assessment of PPS effectiveness for the most-vulnerable scenario for the given threat

7.2.2.1   Adversary Sequence Diagram

An ASD will be constructed for each critical asset included in the strategy. The ASDs will be used to model adversary paths to the critical asset, to derive the most-vulnerable adversary scenarios, and later, to support the risk reduction (system upgrade) function.

The ASD models the PPS at a facility. It identifies paths that adversaries can follow to accomplish the undesired event. An ASD can be used to model all possible adversary paths through a facility. ASDs for buildings may only have one or two layers of protection, but they are helpful tools. They help prevent overlooking possible adversary paths and, when considering protection system upgrades, ASDs help in the selection of upgrades that affect the largest number of adversary paths and can help to ensure that all adversary paths are addressed. For example, suppose that the undesired event is to interrupt or disrupt the information system by attacking the control system operations. Figure 7.1 shows a sample building with two representative physical paths that adversaries might take to damage or sabotage the controls (critical asset) inside the control room. Cyber-attack scenarios will be addressed in the next section.

There are three basic steps in creating an ASD for a specific building. These include:

1. Model the facility by separating it into adjacent physical areas.
2. Define the system features between the adjacent areas.
3. Construct the ASD.

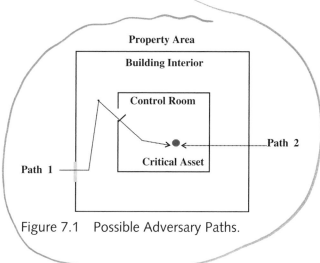

Figure 7.1   Possible Adversary Paths.

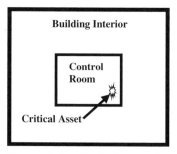

Figure 7.2   Basic Areas at the Example Building.

OffSite
Property Area
Building Interior
Control Room
Critical Asset

Figure 7.3   Adjacent Physical Areas for the Example Building.

7.2.2.2   Physical Areas

The ASD models a facility by separating it into adjacent physical areas. Figure 7.2 is a facility sketch of the example building.

Figure 7.3 describes the adjacent physical areas of the example building. The ASD represents areas by rectangles.

7.2.2.3   Path Elements and Protection System Features

The ASD models a PPS by identifying protection layers between the adjacent areas (see Figure 7.4).

Each protection layer consists of a number of system features. The types of system features used in an ASD include:

- DOOR – Doorway
- DUCT – Duct

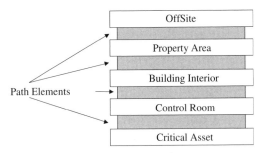

Figure 7.4   Path Elements between
Adjacent Areas.

- FENCE – Fence line
- GATE – Gateway (could be pedestrian or vehicle)
- TASK – Task at critical asset
- PORTAL – Series of two barriers with area between (could be gates or doors)
- SURFACE – Could be wall, roof, floor
- TUNNEL
- WINDOW

The basic ASD as it has been developed so far is given in Figure 7.5. The adversary attempts to sequentially defeat a feature in each protection layer as he traverses a path through the facility to the critical asset. The ASD represents all of the realistic physical paths that an adversary might take to reach a critical asset. For sabotage

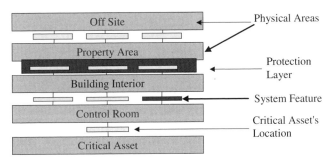

Figure 7.5   ASD Concept.

analysis, only the entry paths would be evaluated, and the system features would be assumed to be traversed in only one direction. For theft analysis, the ASD shown should be considered to be traversed twice – on entry to the critical asset and on exit from the critical asset. Appendix B provides further information on special cases that might be encountered while developing ASDs.

7.2.2.4   Detection and Delay Features

During facility characterization, physical protection features were described, located, and any weaknesses in installation, maintenance, or testing were noted. The next step in analysis is to associate these physical protection features with the path elements of the ASD. Specifically, *detection* and *delay* features of the PPS that are associated with each path element should be annotated. For example:

Identification	Path Element	Detection Features	Delay Features
Personnel door to Control Room	DOOR	Door switch sensor Magnetic stripe card reader	Electronic lock controlled by card reader
		No assessment	Solid wood door

Detection (with assessment) is the discovery of adversary action and includes sensing covert or overt actions. In order to discover an adversarial action, a sensor (either equipment or personnel) must react to an abnormal occurrence and initiate an alarm. Assessment is necessary for effective detection. The information from the sensor and assessment subsystem must be reported and displayed so that someone can ultimately assess the information and determine if the alarm is valid (adversary) or invalid (false alarm or nuisance alarm).

Methods of detection include a wide range of technologies and personnel. Interior sensors can be active or passive (active sensors employ both a transmitter and receiver; passive sensors use only a receiver to sense an intrusion), covert or visible, volumetric or line detection and can be applied to detect boundary penetration, interior motion, or proximity to a critical asset. Exterior sensors can be passive or active, covert or visible, line-of-sight or line detection and can be buried line, fence-associated, or freestanding. Security personnel at fixed posts or on patrol serve a vital role in detection of an intrusion. Personnel can contribute to detection if they are trained in security concerns and have a means to alert authorities in the event of a problem. Assessment can be achieved by a video system displayed at an alarm station or by human observers. Entry control systems allow entry/exit of authorized persons and material, prevent the entry of unauthorized persons, weapons, explosives, or other contraband, and prevent the unauthorized exit of valuable or protected assets. Mary Lynn Garcia, in *Vulnerability of Physical Protection Systems,* describes sensors and components in detail, as well as discusses vulnerability testing for a spectrum of detection equipment.

Entry control, in that it includes locks, may also be considered a delay factor in some cases. Entry control to various layers of the system should be designed to filter and reduce population that has access to the critical asset. Only those who need direct access to the critical asset should be allowed through the final entry control point. Searching for contraband, such as metal (possible weapons or tools) and explosives (possible bombs or breaching charges), is required for high-security areas. This may be accomplished by using metal detectors, x-raying (for packages), and explosive detectors. Personnel also can accomplish contraband detection through physical package searches.

Delay is any physical protection feature that impedes adversary progress. Delay can be incurred by fixed barriers (e.g., doors, vaults, locks). Personnel can be considered an element of delay if they can cause the adversary to be delayed in any way. The time to defeat an obstacle will be dependent on the adversary's capabilities. For example, if the adversary only has a small amount of explosives available, and the delay feature requires a large amount to damage/destroy the barrier, the system will be effective. Note the importance of the detection function in the application of delay. If the adversary can make repeated attempts on a delay feature without intervention, they will eventually defeat the delay element. It is imperative that delay be preceded by a reasonable likelihood of detection earlier in the adversary path.

7.2.2.5 Most-Vulnerable Scenario

The most-vulnerable scenario is an expansion and refinement statement of the most-vulnerable adversary strategy derived earlier in the analysis. The scenario defines the actual path elements that the adversary would traverse and the tactics that he or she would use to reach the critical asset and cause the undesired event. Using the list of path elements annotated with detection and delay features, the analysis team, using expert opinion, should select the path that would optimize the success likelihood for the adversary. Feature weaknesses and gaps, site operating conditions, and possible adversary defeat methods and tactics should be considered in the selection. The most-vulnerable scenario will be used in the next section to estimate the PPS's effectiveness. Software tools like Estimate of Adversary Sequence Interruption (EASI), Systematic Assessment of Vulnerability to Intrusion (SAVI), and Analytic System and Software for Evaluating Safeguards and Security (ASSESS) have been used to derive the most-vulnerable adversary scenario at facilities.

For the example facility, a most-vulnerable scenario for a physical attack might be: *Adversary enters the building through the personnel doors that are unlocked during normal working hours, traverses the building interior to the control room. When control room is unoccupied, adversary forces control room door open, sets explosives/timer on computer equipment, and exits.* The corresponding path elements from the ASD include:

- DOOR – Personnel entrance to building
- AREA – Building Interior – traversal
- DOOR – Control Room door
- AREA – Control Room – traversal
- TASK – Set explosives/timer

7.2.2.6   Estimating Physical Protection System Effectiveness

PPS effectiveness is the measure of the ability of the PPS to meet its specified protection objectives, to prevent the specified undesired events. The most-vulnerable scenario and associated system features developed in the previous discussions will be used to estimate protection system effectiveness for each undesired event. The assumption of the analysis process is that the effectiveness of the PPS is only as good as the protection that it provides against the most-vulnerable scenario.

An effective PPS must be able to detect the adversary early, delay the adversary long enough for the security response force to arrive, and neutralize the adversary before the undesired event is accomplished. In particular, an effective protection system demonstrates effective detection, delay, and response. These physical protection functions (detection, delay, and response) must be integrated to ensure that the adversarial threat is neutralized before their mission is accomplished. Detection is the sensing of an adversarial action and the assessment that it is a valid alarm; delay is any protection feature that impedes the adversary's progress; response comprises actions taken by the security police force (police force or law enforcement officers) to prevent adversarial success. The security response must be notified in a timely and reliable manner,

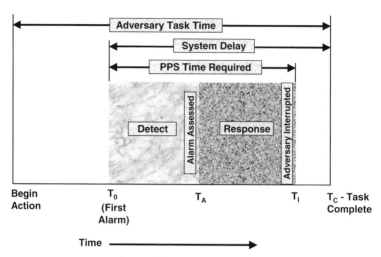

Figure 7.6   Interrelationships of PPS Functions.

must arrive in time, and must be physically capable of neutralizing the adversarial action before the undesired event is achieved.

Relationships of PPS Functions – The diagram below (see Figure 7.6) shows the relationships between the adversary's task time and the time required for the PPS to do its job. The total time required for the adversaries to accomplish their goal has been labeled Adversary Task Time; it is dependent upon the delay provided by the PPS. The adversary may begin the task at some time before the first alarm occurs (T_0). The adversary task time is shown before T_0 because delay is not effective before detection. After the alarm, the information must be reported and assessed to determine if the alarm is valid. The time at which the alarm is assessed to be valid is T_A, and at this time, the location of the alarm must be communicated to the members of the response force. Further time is then required for the response force to respond in adequate numbers and with adequate equipment to interrupt the adversarial actions. The time at which the response force interrupts the adversary is T_I, and adversary task time completion is T_C. For the PPS to accomplish its objective, T_I must occur before T_C. From this diagram, it is obvious that a PPS performs better if detection is as early in the timeline as possible and delay elements are near the critical asset and location.

System effectiveness analysis can be performed by simply checking for required features of a protection system, such as intrusion detection, entry control, access delay, response communications, and a response force. However, a protection system based on required features cannot be expected to lead to a high-performance system unless those features, when implemented together, are sufficient to ensure adequate levels of protection. Sophisticated analysis and evaluation techniques can be used to estimate the minimum performance levels achieved by a protection system. Computer codes such as EASI, SAVI, ASSESS and Joint Combat and Tactical Simulation (JCATS) can be used to estimate a protection system's effectiveness. For applications here, protection system effectiveness schemes will be discussed for a simple protection system and a more complex system.

7.2.2.7 Simple Physical Protection System

A simple protection system can be briefly described as a protection system with a small number of protection features and only a small number of path elements to protect against adversarial actions. In most cases for a simple system, judgment of system effectiveness can be made by inspection. If one or more of the physical protection functions (detection, delay or response) are absent, lacking, or judged to be grossly ineffective, system effectiveness is low. Because all three functions are required, the generalization is that the protection system is only as good as the weakest link. For example:

Detection	*Delay*	*Response*
Door switch sensors	*Hardened doors and walls*	*None: No arrangements made with local law enforcement*

Detection	Delay	Response
Magnetic swipe card reader	*Electronic locks*	
No assessment	*Moderate area traversal time*	
Low effectiveness	Medium effectiveness	Low effectiveness
Minimum of Detection, Delay, and Response: Low effectiveness		

Protection system effectiveness would be estimated at *Low* effectiveness by selecting the minimum level of effectiveness for the protection functions of detection, delay, and response.

7.2.2.8 Complex Physical Protection System

If the PPS is more complex (i.e., there are various protection features for each system function), it may be more difficult to judge if the features demonstrate a high level of performance for detection, delay, and response features and whether or not the system would be expected to detect the adversary early enough, provide enough delay time to ensure that the response could arrive and then neutralize the adversarial action before the undesired event was achieved. In the absence of software tools to estimate a likelihood of adversary interruption before achievement of the undesired event, a first-order tool will be discussed to provide a qualitative estimate of protection system effectiveness. Consider a more complex facility and PPS shown in Figure 7.7.

Layer 1 of path elements between Off Site and the Property Area would include the GATE (pedestrian), the FENCE, and the GATE (vehicle); Layer 2 of path elements between the Property Area and the Building Interior would include DOOR (either one of two personnel doors), the SURFACE (walls, roof, floor), and the

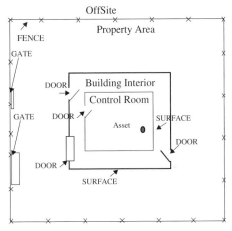

Figure 7.7   Example Facility with
Complex Protection System.

DOOR (vehicle); Layer 3 of path elements between the Building
Interior and the Control Room would include SURFACE (walls,
ceiling, floor), or the DOOR (personnel). Note that if path elements
have identical protection features, they can be modeled once on the
ASD. For example, if all personnel doors into the building have the
same construction, locks, sensors, and assessment, only one DOOR
will appear on the ASD. The corresponding ASD might look like
Figure 7.8.

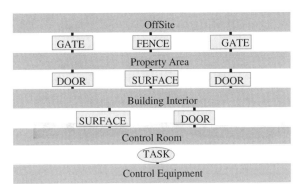

Figure 7.8   ASD for Example Facility with
Complex Protection System.

Adversary
sequence
diagram

A table can be constructed of the path elements for the ASD that is annotated with detection and delay features. See Table 7.1. The analysis team can use this table together with other information collected during facility characterization and expert judgment to derive a most-vulnerable adversary scenario.

The path elements listed on the rightmost column are those judged by the team to be associated with the most-vulnerable path for a specific adversary. The next step is to assess if the protection features associated with these path elements could perform together to accomplish the protection objective of preventing the loss of the information system by a physical attack. One of the sophisticated software tools would estimate a probability of interrupting the adversary using a path algorithm that would address different adversary tactics and tools.

The basic task is to estimate whether or not the adversary would be expected to be interrupted before the undesired event was achieved. For our purposes here, a crude time line is created for the adversary to complete the scenario judged by the team to be most-vulnerable. The adversary task time is accumulated from the path element that is judged to have a detection level of *Medium* or higher up to and including the task to accomplish the undesired event. Then the adversary task time is compared to the response force time. If the adversary task time (after detection has occurred) is shorter than the response force time, the adversary is not expected to be interrupted before the undesired event occurs; if the adversary task time is significantly longer than the response force time, interruption should occur.

"Appendix C, System Effectiveness Worksheets" contains worksheets for path elements to estimate relative/qualitative values of detection for a spectrum of detection features and delay time estimates for various barriers and traversal times, assuming a moderate adversarial threat with basic hand and power tools and

Table 7.1 Detection and Delay Features for Path Elements for Example

	Path Element	Detection	Delay	Selected Path Element
Layer 1	Vehicle Gate	*No features*	*Normally locked Wrought iron gate*	Pedestrian Gate
	Fence	*No features*	*5 ft wrought iron*	
	Pedestrian Gate	*No features*	*Always open*	
	Area: Traversal Distance – *100 ft*			Property Area
Layer 2	Vehicle Door	*No features*	*Metal roll up door Locked off hours*	Pedestrian Door
	Surface	*Personnel during working hours*	*Reinforced block walls*	
	Pedestrian Door	*Receptionist during working hours Door alarmed off hours*	*Tempered glass door Key locked off hours*	
	Area: Traversal Distance – *50 ft*			Building Interior
	Surface (control room)	*Control room manned 24/7*	*Framed sheetrock walls*	Control Room Door

Table 7.1 (continued)

	Path Element	Detection	Delay	Selected Path Element
Layer 3	Door	Badger reader Door switch alarm Control room personnel 24/7	Hollow-core metal door Electromag-netic stripe lock	
	Area: Traversal Distance – 5 ft			Control Room
	Task	Control room manned 24/7	No delay features	Task

explosives. The process is to complete worksheets to estimate detection level and delay time for each path element and area of the most-vulnerable scenario. Then the results can be summarized to assess whether or not the adversary would be expected to be interrupted and the undesired event prevented. Table 7.2 summarizes the results for the example's most-vulnerable scenario. Note that adversary delay time does not accumulate until *after* a *Medium* or higher level of detection. The derivation of these example results is also included in Appendix C.

7.2.2.9 Physical Protection Vulnerabilities

In summary, the effectiveness of the PPS is judged to be at the *Low* level for the given adversarial threat. Further, each *Low* level assessment for path elements indicates weakness that leads to a specific vulnerability. The specific weakness or deficiency that drives the judgment to be *Low* instead of *Medium* or *High* is a site-specific vulnerability.

physical
protection
System

Table 7.2 Summary Assessment Results for Example Scenario

Path Elements for Most Vulnerable Scenario	Detection Level	Delay Time (seconds)
Gate (pedestrian)	Low	0
Property Area		7
Door (pedestrian)	Medium	30
Building Interior Area		3.5
Door (pedestrian)	Medium	10
Control Room Area		0
Task at Target	Medium	15
Delay Time after Detection		A: 28.5 seconds
Response Time		B: 300 seconds
Estimated System Effectiveness Level A < B, System Effectiveness = Low A ~ B, System Effectiveness = Medium A > B, System Effectiveness = High		System Effectiveness: Low

For the example, the pedestrian gate on the perimeter is judged to be *Low* detection effectiveness because there are no detection features to detect an intruder during working hours or off hours. The site-specific vulnerability in the PPS at the gate is the lack of detection features. There are no access controls, contraband detection, or intrusion detection features.

Identification of specific vulnerabilities is important for contingency planning and for system upgrades to reduce security risk. For contingency planning, during heightened threat conditions, additional security features can be implemented, even temporarily (if not affordable permanently) to protect the facility. In order to reduce risk, the first approach is to increase protection system effectiveness. Addressing the site-specific vulnerabilities will increase protection system effectiveness, and so, reduce relative security risk.

7.2.3   Cyber-Protection System Effectiveness

Some critical assets may be susceptible to malevolent attack by a physical and cyber-attack or maybe just a cyber-attack. This section will discuss another first-order method to assess the effectiveness of the cyber-protection system for a given critical asset. The basic protection objective for cyber-protection systems is to protect information and information systems. Specifically, the protection objective is to preserve the following three properties for data:

- Confidentiality
- Integrity
- Availability

Confidentiality requires that information not be made available to unauthorized individuals, entities, or processes. Confidentiality requirements vary greatly, depending upon the category of information. There are no confidentiality concerns for nonsensitive information, but there are stringent needs for maintaining the confidentiality of highly sensitive information – for example, critical process control assets or control communications.

Integrity requires that information not be altered or destroyed in an unauthorized manner. Although integrity concerns can vary with the information in question, there is a need to preserve the integrity of nearly all information; otherwise, there would be no value in maintaining the information. In the case of mission-critical information or information affecting safety, the level of concern regarding integrity can be quite high.

Availability requires that information be accessible and usable on demand by an authorized entity. The level of concern regarding availability can vary greatly, depending upon the information and the uses to which it is put, and costs of implementation and operations generally increase as availability requirements increase. In the case of some mission-critical applications, it may be necessary and prudent to build redundancy into the system at considerable expense.

Analogously to the PPS assessment method, the cyber-protection system assessment method includes:

- Development of a cyber-path-diagram for the critical asset
- Association of authentication, authorization, and audit features for electronic paths to the critical asset
- Assessment of cyber-protection system effectiveness for the critical asset

7.2.3.1 Cyber Path Diagram

All cyber-paths that link to a critical cyber-asset must be protected. Specifically, each cyber-link to the critical asset must be subject to the authentication, authorization, and audit functions. Analogous to the ASDs for physical paths to a critical asset, cyber-path diagrams can be constructed to describe cyber-paths to the critical cyber-asset.

The first step is to identify the electronic security boundaries between the exterior of the system and the critical cyber-asset. Normally cyber-systems have an exterior electronic boundary and one or more interior boundaries. Cyber-protection features are deployed at these boundaries.

The next step is to identify all of the access points to the system. Systems can be accessed via modems (located on- or off-site), the Internet, control room, alternate access points in the facility, communication links, or by the downloading of software.

The electronic links between boundaries must be identified. These links can be formed by other noncritical cyber-assets or direct communication links.

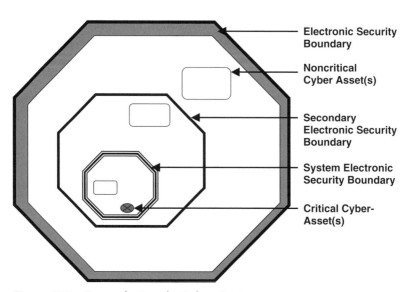

Figure 7.9   Example Simple Cyber-System.

Figure 7.9 depicts a very simple cyber-system to demonstrate the concept. The system has a critical asset within an electronic security boundary at the perimeter, a secondary electronic security boundary, and a system-level electronic security boundary, with various other noncritical cyber-assets. The numerous communication links that exist between entities are not shown on the chart.

Figure 7.10 includes a cyber-path-diagram for the critical asset of the example simple cyber-system. This diagram can be used to consider the numerous cyber-paths from the access points to the critical cyber-asset.

7.2.3.2   Cyber-Protection Functions

Much like an effective PPS demonstrating high performance for the three functions of detection, delay, response, and the integration of these functions, an effective cyber-protection system demonstrates high performance for three basic cyber-security functions and their integration. These functions are used to ensure the properties

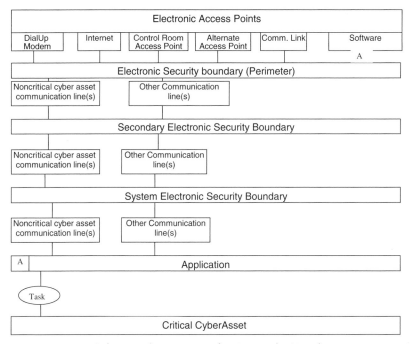

Figure 7.10   Cyber-Path-Diagram for Example Simple Cyber-System.

of confidentiality, integrity, and availability. The three functions include:

- Authentication
- Authorization
- Audit

7.2.3.3   Authentication

Authentication is the process of establishing the validity of a claimed identity. User authentication is the process of associating a computer identity with a human being. This may be done using mechanisms that fall into three basic categories: (1) something the individual knows, (2) something the individual has, and/or (3) something the individual is. Once a user is authenticated, he or she is generally issued credentials that are associated with computer processes acting in the user's behalf. User authentication is critical to the overall security of a system or network, because if one user obtains (maliciously or otherwise) another user's credentials, then he or she can access any information that user is permitted to access. Two-factor authentication means authentication requiring two (or more) of the above factors. Two-factor authentication is stronger than authentication based upon a single factor, especially when that single factor is a password.

The most frequently used authentication mechanism is the password, which is something the individual knows. Passwords are more exploitable than most other authentication mechanisms. If a password is stolen or compromised, the original owner retains use of it, while at the same time another user can use it for a considerable period of time without the owner's knowledge. In order to reduce the risk of compromised passwords, encryption techniques are frequently used to protect passwords when they are stored on a system and when they are transmitted over a network; however, this does not protect the passwords against keystroke capture at the client machine.

Smart cards or tokens represent "something the user has," and their use has become more prevalent in recent years. Smart cards and tokens can be divided into two major subcategories: (1) smart cards/tokens that connect electronically to the user's system and (2) one-time-password tokens that interface to the user only via a touch pad and display. Smart cards can potentially be compromised via a network attack, although this is much more difficult than compromising a reusable password. One-time-password tokens are not as subject to misuse, because they require human interaction upon every use, but they are therefore considerably less convenient. On the other hand, smart cards are more convenient, because they require less interaction by the user, and they also support encryption and digital signature functions, as well as authentication.

The biggest advantage of smart cards and tokens is that if they are lost or stolen, the owner is immediately aware of the fact, since he/she loses access. If the loss is reported, the device can immediately be disabled at the authentication server to prevent further use.

Biometric authentication is based upon "something the user is." Biometric technology has not yet been widely accepted, because of both its cost and the difficulty of reaching an acceptable level of false positives and false negatives. Fingerprint recognition is currently the most popular and socially acceptable biometric technology, and the cost and accuracy of fingerprint readers has dropped dramatically. The use of this technology requires fingerprint readers and software to be deployed to the client systems.

Because of their role in cyber-security, all process control network authentication servers will be afforded the maximum protection practical. In order to maintain the confidentiality and integrity of these central authentication services, it is imperative that the number of persons with privileged access (e.g., root or administrator) to these services be kept at a minimum and that these employees be appropriately screened.

7.2.3.4   Authorization

Authorization is the process of determining what actions an entity is allowed to perform with respect to a given object. Authorization for access to systems and applications must be granted by management. Authorization for access to information on systems must be controlled so that only authorized users can access specified information objects (e.g., files, data base records, web pages) based upon their authenticated identity.

7.2.3.5   Audit

Auditing is the process of recording the actions or attempted actions performed by an entity within a computer system or network.

The intrusion detection system supports the audit function. The major components of a cyber-intrusion detection system include the review of traffic data; scanners to detect any unusual occurrences, including any suspect ports or modems; virus protection; and monitors for access control.

All operating systems and applications services must log security significant events. Where possible, these events should be recorded in the system log in order to facilitate access to these events by centralized audit log analysis tools. The logs gathered on client workstations will not normally be examined, except in the case of an incident investigation.

The primary tools used to detect vulnerabilities in operating systems and network applications are network vulnerability scanners. The most significant vulnerabilities that exist on a system are generally the ones that are visible from outside the system and that make it vulnerable to network attack. For this reason, emphasis is placed on network vulnerability scanners. It is important that the vulnerability scanning, analysis, and reporting process be automated to the extent possible.

Operating systems and applications must be securely configured with all applicable patches, and these patches must be kept up to date as new vulnerabilities are discovered. Systems that are externally accessible must be updated immediately with security significant patches. Systems that are not directly accessible from the outside must still be patched, but the time frame will vary according to the seriousness of the vulnerability. Vulnerability analysis tools should be used to verify that systems have the required patches.

One of the most convenient avenues of attack against networks is through the introduction of malicious code onto machines. Virus protection is used to support the detection of malicious code. Any software packages that are added should be carefully reviewed and tested and connection to the web should be protected against or prohibited.

The audit function includes access control monitoring. There is a complementary relationship between firewalls and intrusion detection systems. Firewalls block undesired network traffic and permit desired traffic. The cyber-intrusion detection system inspects both blocked and permitted traffic for suspect patterns.

7.2.3.6 Integration of Cyber-Functions

Each of the cyber-functions of authentication, authorization, and audit must be performed at a high level, and the functions must be integrated. The authentication and authorization functions both provide data to the audit function where it is analyzed for evidence of malicious activity. Firewalls and encryption support all three cyber-functions as well as the protection of the communication links used among the functions.

7.2.3.6.1 *Effective Cyber-Protection System* An effective cyber-protection system provides graded protection, namely security measures must be commensurate with the sensitivity of the information contained in that system. If a critical cyber-asset can be maliciously compromised to cause a high-consequence undesired event, a high level of protection must be afforded it. Cyber-security measures are implemented at the network, system, and application level primarily to protect the information contained therein, although in a few cases security measures are implemented to prevent unauthorized access to high-value systems themselves, such as any critical cyber-assets or control communication link.

Cyber-protection system effectiveness is the measure of the ability of the cyber-protection system to *prevent* the undesired event. All of the cyber-paths to the critical cyber-asset must be protected. The assumption of the analysis process is that the effectiveness of the cyber-protection system is only as good as the protection that it provides for all of the electronic paths to the critical cyber-asset. The process for estimating cyber-protection system effectiveness has three basic parts:

1. List features of the cyber-protection system that provide authentication, authorization, auditing, and system integration for the critical asset.
2. Estimate cyber-protection system effectiveness for each asset by assessing the performance level of protection system features for authentication, authorization, audit, and system integration from Table 7.3.

Table 7.3 was established by expert opinion provided by cyber-analysts.

Continuing with our example, if our critical cyber-asset is the electronic control system, the cyber-protection system assessment might look like the results in Figure 7.11. Authentication is judged to be *Low* because user defined passwords are considered low effectiveness; authorization is *Low* because all employees and contractors have access to critical cyber-assets; perhaps audit performance is judged to be *Medium* in effectiveness and not *High* because data reviews are periodic and not timely; the integration function is judged to be *Low* because there are no firewalls or encryption features to integrate the system performance.

7.2.3.7   Cyber-Protection Vulnerabilities

In summary, the effectiveness of the cyber-protection system is judged to be at the *Low* level for the given moderate adversarial threat. Further, each *Low* level assessment for cyber-functions indicates a specific vulnerability. The specific weakness or deficiency that drives the function to be judged as *Low* instead of *Medium* or *High* represents a system vulnerability. For the example, weak passwords (user-defined) represent a site-specific vulnerability as does the lack of cyber-system-integration features.

Table 7.3 Relative Cyber-Protection System Effectiveness

	Low Effectiveness	*Medium Effectiveness*	*High Effectiveness*
Authentication	No features or weak password[a]	Strong password[b]	two-factor
Authorization	No features or permissions based upon coarse groupings, e.g., "any employee"	Permissions based upon project-based groups or roles	Permissions based upon project-based groups or roles; other user attributes, and/or authentication trust level (e.g., two-factor)
Audit	No features	Required, retained for X months, analyzed if incident occurs	Required, Retained for X months, analyzed periodically for evidence of unauthorized activity
Integration	No firewalls and/or No encryption	Some firewalls and some encryption	Firewalls and encryption for all paths

[a] Weak password: No requirements for length or type of characters (relatively easy to defeat).
[b] Strong passwords have requirements for length, use of characters (letters, capitals, numbers; are relatively harder to defeat).

1. List cyber protection system features for paths			
Authentication Passwords – User defined	**Authorization** All employees and contractors have standard authorization. Only system administrators can access corporate system.	**Audit** Intrusion detection system at perimeter electronic boundary Scanners Virus protection Access control monitoring Random traffic data review	**Integration** No features

2. Estimate cyberprotection system effectiveness for paths to asset: L

Cyberprotection system (circle one)

Authentication performance level: L M H

Authorization performance level: L M H

Audit performance level: L M H

System Integration: L M H

Minimum value of above   L M H

Minimum level = cyberprotection system effectiveness for asset ___L___

Figure 7.11   Example Cyber-Protection System Effectiveness Assessment.

Identification of system vulnerabilities is important for system upgrades to reduce security risk. Addressing the system's vulnerabilities will increase the protection system's effectiveness, and so reduce relative security risk.

7.3   SUMMARY

In this chapter, protection system effectiveness has been discussed and demonstrated. The outputs of system effectiveness assessment are the estimation of PPS effectiveness, cyber-protection system effectiveness, if appropriate, and identification of site-specific vulnerabilities in the protection system.

System effectiveness assessment uses the site-specific fault tree, the PPS description, and the cyber-protection system description to assess whether or not the system meets the specified protection objectives for the defined threat description.

The system effectiveness assessment methods described in this chapter can be used for both outsider threats and an insider threat.

A protection system to mitigate the insider threat faces challenges because of the knowledge, access, and authorization afforded the insider. Protection from the insider adversary is discussed separately in "Appendix D, Insider Threat."

7.4 REFERENCES

1. Biringer, Betty, "Risk Assessment Method for Electric Power Transmission," presented at Carnahan Conference on Security Technology, sponsored by IEEE, Albuquerque, NM, October 2004.
2. Brown, C. Douglas, Sandia National Laboratories "Cyber Security Architecture for Unclassified Computer Environments," Sandia National Laboratories, September 30, 2004.
3. Garcia, Mary Lynn, *The Design and Evaluation of Physical Protection Systems*, Butterworth-Heinemann, Boston MA, 2001.
4. *International Training Course for Nuclear Facilities and Materials – Volume I. Determining Physical Protection System Objectives*, Sandia National Laboratories and the International Atomic Energy Agency, 2004.
5. North American Electric Reliability Council, "Urgent Action Cyber Security Standard, Standard CIP-002-1," Draft, May 9, 2005, http://www.nerc.com/~filez/standards/Cyber-Security-Permanent.html.
6. Paulus, W. K., "Generic Physical Protection Logic Trees, SAND79-1382," Sandia National Laboratories, Albuquerque, New Mexico 87185, October 1981.
7. Sandia National Laboratories, *Analytic System and Software for Evaluating Safeguards and Security (ASSESS) User's Guide*, March 1993.
8. *Sandia National Laboratories Security Risk Assessment Methodologies*, http://www.sandia.gov/ram.
9. "Understanding Risk in a Changing World," short course taught at Society of Women Engineers National Conference, October 16, 2004, Milwaukee, WI.
10. Vesely, W. E., Goldberg, F. F. Roberts, N. H., and Haasi, D. F. *Fault Tree Handbook*, NUREG-0492, Systems and Reliability Research, Office of Nuclear Regulatory Research, U.S. Nuclear

Regulatory Commission, Washington, DC 200555, January 1981. (Available from GPO Sales Program, Division of 20555 and National Technical Information Service, Springfield, VA.

7.5   EXERCISES

1. List the information developed in previous chapters that is used extensively in the protection system effectiveness assessment.
2. How is the site-specific fault tree used to identify adversary strategies and eventually to identify the most-vulnerable scenario?
3. Under what conditions are both physical and cyber-protection system assessments required?
4. Describe the general process to determine the most-vulnerable strategy and scenario. What site information is used in the determination?
5. Consider the following hypothetical facility:

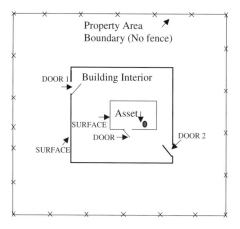

a. Sketch the ASD for sabotage of the asset.
b. Assume that the most-vulnerable scenario includes the path elements: DOOR 2, Building Interior Area, DOOR, TASK (sabotage asset).
  (i) If there are no detection features, what level is the physical protection system's effectiveness? Why?
  (ii) If there are no detection features, but delay and response are at the *High* level, what level is physical protection system effectiveness? Why?

    (iii) If detection and delay are at the *High* level and there are no response features, what is the level of the physical protection system's effectiveness? Why?

  c. If there are various features for detection and delay and there is a local law enforcement response, list which worksheets might be used to estimate the physical protection system's effectiveness?

  d. Assume the following features for the path elements of the most-vulnerable scenario. Assume that the response is judged to be highly effective, is located nearby, and can reliably arrive within five minutes from the time that the response team gets the call.

Path Element	*Detection*	*Delay*
Pedestrian Door	*Door alarmed off hours*	*Tempered glass door* *Key locked off hours*
Building Interior Area: Traversal Distance – *30 ft*		
Vault Room Door	*Badge reader* *Door switch alarm*	*Steel door* *Combination lock*
Task	*Motion sensors* *Camera assessment*	*No delay features*

    (i) Estimate the PPS effectiveness level.

    (ii) List site-specific vulnerabilities in the PPS.

6. Assume that the following graph describes the cyber-protection system for critical cyber-assets and that the system is accessible via the Internet, dial-up modems, and the control room.

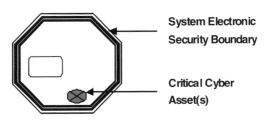

System Electronic Security Boundary

Critical Cyber Asset(s)

a. Sketch the cyber-path-diagram for the system.
b. Assume that the Electronic Security Boundary includes the following cyber-protection features:

Authentication	Authorization	Audit	Integration
Passwords required – machine-generated	Short list of employees/contractors who are allowed to access system	Required, system analyzed on prescribed schedule for unauthorized activity	Series of firewalls

c. Estimate the cyber-protection system effectiveness level.
d. List any site-specific vulnerabilities of the cyber-protection system.

7. Assume that a critical asset is susceptible to physical attack and cyber-attack and that either type of attack will cause the same level of consequences. If the PPS effectiveness is estimated to be *Medium* and the cyber-protection system effectiveness is estimated to be *Low*, what is the overall level of protection system effectiveness? Give reasons for your response.

8. Discuss the value of site-specific vulnerabilities in the protection system that have been derived by a systematic analysis.

Chapter 8

Estimating Security Risk

8.1 INTRODUCTION

At this point, processes have been provided to estimate the three parameters to estimate security risk: likelihood of adversary attack, system ineffectiveness, and the consequences of adversary success. What is important to the mission of the facility has been identified; specifically, the undesired security events that would interrupt the mission, the consequences associated with the event, and the targets that must be protected to prevent the undesired events. The adversarial threat spectrum, who might attempt the undesired event(s), has been described in as much detail as possible. A system effectiveness analysis has been completed for both the PPS and the cyber-protection system to determine how well the current protection system protects against the adversarial threat spectrum for the undesired events and to identify site-specific vulnerabilities. The next step is to combine the three security risk parameters (likelihood of adversary attack, system ineffectiveness, and the consequences of adversary success) to estimate security risk.

8.2 ESTIMATING SECURITY RISK

Security risk managers need a "measurement scale" to help them use the information that they have, to make the most logical

121

business decisions to manage security risk from malevolent acts. The security risk value estimated in this chapter is a qualitative estimate of security risk. The purpose of this risk value is to provide a reference point or "measure" for the security risk associated with the baseline protection system. This "measurement scale" is particularly important to risk managers because it helps them understand the level of current security risk, and it provides a reference point for evaluating and comparing other security risks.

8.2.1 Conditional Risk

Conditional security risk does not include the initiating event (adversary decides to attack) and focuses on the likelihood of adversary success in the attack (system ineffectiveness – the complement of protection system effectiveness) and the consequences resulting from the attack. Historically, conditional security risk has been used when there is not enough information to estimate attack likelihood and/or when consequences are so extremely *High* (unacceptable), that risk analysts do not bother with estimating the likelihood of the attack. Of the three security risk parameters, likelihood of adversary attack is the most uncertain because it is difficult to estimate and is the most subjective of the parameters.

From the security risk equation, conditional risk is expressed as:

$$\text{Risk} = (1 - P_E) * C$$

$$P_E = \text{System effectiveness}$$

$$(1 - P_E) = \text{System in-effectiveness}$$

$$C = \text{Consequence}$$

8.2.2 Relative Risk

The risk value estimated in this chapter is the relative security risk that is qualitative in nature. It is used to provide a reference level of risk and for comparison purposes. The risk level is expressed as

Very High, High, Medium, Low, and *Very Low*. The temptation is to associate numbers with the levels and derive a quantitative value for the risk level. The process can be used as long as the results are qualified in terms of the accuracy of the resultant quantitative value. The result is a point estimate, at best, and should not be used as an absolute value; instead it should be used only to establish relative ranking.

For our purposes here, qualitative levels of the three risk parameters (likelihood of adversary attack, system in-effectiveness, and consequence of attack) are combined utilizing a combination of logic and expert judgment model. Table 8.1 provides an example table, combining the risk parameters to estimate relative security risk. For example, for a given scenario to cause the undesired event, if

Table 8.1 Estimating Relative Security Risk

P_A (Likelihood of Attack)	$1 - P_E$ (System Ineffectiveness)	C (Consequence)	R (Relative Security Risk)
L	L	L	L
L	L	M	L
L	L	H	L
L	M	L	L
L	H	L	L
L	M	M	M
L	H	M	M
L	M	H	H
L	H	H	H
M	L	L	L
M	L	M	L
M	L	H	L

Table 8.1 *(continued)*

P_A (Likelihood of Attack)	$1 - P_E$ (System Ineffectiveness)	C (Consequence)	R (Relative Security Risk)
M	M	L	L
M	H	L	L
M	M	M	M
M	H	M	M
M	M	H	H
M	H	H	H
H	L	L	L
H	L	M	L
H	L	H	L
H	M	L	L
H	H	L	L
H	M	M	M
H	H	M	M
H	M	H	H
H	H	H	H

L = Low
M = Medium
H = High

the threat potential or likelihood of an adversary attack is estimated to be *Medium* (M), the system ineffectiveness is estimated
to be *Low* (L), and the consequences of the attack are estimated
to be *Medium* (M), the relative security risk is estimated to be
Low.

8.3 SUMMARY

In this chapter, the three parameters of the security risk equation – likelihood of adversary attack, system in-effectiveness, and consequence – are combined to estimate relative security risk. The overriding caution is that the estimated value of security risk is relative in nature and is a qualitative point estimate, at best. The value is not absolute, and any further degree of numerical accuracy should not be implied.

The estimated security risk value does provide risk managers with a valuable "measure" of risk to malevolent attack that can be used to make risk management decisions.

8.4 REFERENCES

1. *Sandia National Laboratories Security Risk Assessment Methodologies*, http://www.sandia.gov/ram.
2. Paulus, William and Matalucci, Rudy, "Risk Matrix Table," Sandia National Laboratories, Albuquerque, NM, June 2001.

8.5 EXERCISES

1. Define conditional risk. When is conditional risk preferred?
2. Define what is meant by relative security risk.
3. Discuss the benefits of having a "measurable" parameter for security risk level.
4. Discuss how security risk values *should* and *should not* be interpreted.
5. Describe how and why the assessment team might decide to modify the table for Estimating Relative Security Risk (like Table 8.1).

Chapter 9

Risk Reduction Strategies

9.1 INTRODUCTION

At this stage in the security risk management process, the perception of the analysis team is that the security risks are above the acceptance level, and the presentation package to management should include analysis of security strategies and mitigation options to reduce risk for the building or facilities that are being reviewed. Most likely the direction has been set for exploring and developing alternative plans that would accomplish and demonstrate risk reduction and ultimately enhance the protection of occupants, property, and mission requirements. This chapter will describe the development of strategies to reduce security risk. The strategies are based on the parameters of the security risk equation: likelihood of adversary attack, system ineffectiveness, and consequence. Logically, in order to reduce security risk, one or more of the three parameters must be reduced. The analysis team will explore and analyze possible strategies to reduce one or more of the risk parameters. Figure 9.1 outlines the basis of security risk reduction strategies.

9.2 STRATEGIES FOR REDUCING LIKELIHOOD OF ATTACK

Reducing the likelihood of adversary attack points toward some type of "deterrence" strategy. Security deterrence is very difficult to

127

Figure 9.1   Basis of Risk Reduction Strategies.

measure. Historically, deterrence appears to be effective for some period of time; the period of time is hard to predict because it is only as long as the time required for the adversary to learn how to defeat the system. The most reliable deterrence is an effective security system.

For lower-level threats or for threats without a specific target, there is some credence in the belief that a well-structured building security system might "deter" an attack or rather redirect it to a more vulnerable neighboring building, if the objectives for the attack are similar in motivation, and the impact on society and the country might be relatively equitable. Of particular interest are the deterrence effects certain security and protective measures can produce if the adversary is less determined, less motivated, and less willing to die for a cause. The less sophisticated adversary might be deterred from an attack if cameras are constantly providing surveillance and are difficult to compromise. Crime-witness methods can also be effective, such as posting reward signs around a building or in the vicinity of its facilities that indicate benefits to the reporting witness if notification is made to any authority about any indication of a criminal activity in the surrounding area or at the building. In addition, surveillance cameras may be useful for forensic purposes and prosecution.

For high-level motivated threats, it is not prudent to base a reduction strategy solely on reducing likelihood of adversary attack. For these reasons, the focus for security risk reduction strategies will be the improvement of security system effectiveness, the mitigation of consequences, or a combination of both security system improvement and consequence mitigation.

9.3   STRATEGIES FOR INCREASING PROTECTION SYSTEM EFFECTIVENESS

In an ideal world, the goal would be for the security system to prevent all of the undesired events, so exploring strategies to increase the protection system's effectiveness is the first risk reduction strategy to be discussed. The obvious starting point for ways to increase the protection system's effectiveness is the list of site-specific vulnerabilities identified during the system effectiveness analysis, Chapter 7, "System Effectiveness." These specific vulnerabilities were a product of the analysis; logically, removing or securing these vulnerabilities would increase protection system effectiveness and, thus, reduce security risk.

9.3.1   Physical Protection System Upgrades

Physical protection system upgrades address the detection, delay, and response functions and their integration. Table 9.1 provides examples of possible system features to be considered for upgrades, listed by protection system function.

9.3.2   Cyber-Protection System Upgrades

Cyber-protection system upgrades address the authentication, authorization, and audit functions and their integration. Table 9.2 provides examples of possible system upgrades by protection function.

9.3.3   Protection System Upgrade Package(s)

A protection system upgrade "package" that addresses all of the site-specific vulnerabilities for the building or facility can then be suggested and assessed with a system effectiveness analysis to ensure

Table 9.1 Potential Physical Protection System Upgrade Features

Detection	Delay	Response	Integration
Intrusion Sensing: *Interior sensors:* Boundary penetration Motion Proximity Personnel *Exterior sensors:* Intrusion Personnel *Access control:* Identity check Authorization Contraband detection *Alarm communication:* Tamper indication Line supervision *Assessment:* CCTV Personnel	Barriers Locks Security personnel Tasks at critical asset	Interruption: Communication to response Deployment time Neutralization	System detects adversary early enough and delays the adversary long enough for the response to arrive

and measure the increase in protection system effectiveness. The specific steps to develop a security system upgrade package to increase protection system effectiveness include:

1. Review of the site-specific vulnerabilities
   a. Protection functions associated with each vulnerability
      (i) Physical (detection, delay, response, integration)
      (ii) Cyber (authentication, authorization, audit, integration)

Table 9.2   Potential Cyber-Protection System Upgrade Features

Authentication	Authorization	Audit	Integration
Something known: Passwords **Something possessed:** Smart cards Tokens **Personal identifier:** Biometric (fingerprint) **Dual-factor:** Any two of above	Determined by management Limits number with access Controlled	Network scanners Intrusion detection Review of traffic data Scans Virus protection Access control monitors	Encryption Firewalls

    b. Specification of the system feature associated with each vulnerability

  2. Suggestion and evaluation of protection system upgrade package(s) to increase protection system effectiveness

A summarization and organization of all of the site-specific vulnerabilities identified by the system effectiveness analysis provides valuable guidance about where to upgrade the protection system. The lists of site-specific vulnerabilities identify the protection function, as well as the specific system feature associated with each vulnerability. Remember that a system feature is one of the path elements from the ASD, namely, a DOOR, DUCT, FENCE, GATE, TASK, PORTAL, SURFACE, TUNNEL, or WINDOW. Upgrade features must be suggested for site-specific vulnerabilities. Depending on the nature of the vulnerability, upgrades must be made to the PPS system and/or the cyber-protection system.

    One or more security system upgrade packages should be developed from the feature upgrades to address all of the identified site-specific vulnerabilities. Packages should be based on the judgment of the analysis team. Upgrade package content may be driven by threat level, ease or timeliness of implementation or site or corporate sensitivities. Most often, three security system upgrade packages are postulated, based on threat level:

1. Package 1 is designed for a lower-level threat (vandal, gang, criminal, extremist)
2. Package 2 is designed for a medium relative threat (Package 1 plus a domestic terrorist)
3. Package 3 is designed for a higher relative threat (Package 2 plus an international terrorist threat plus an insider)

Adversary Sequence Diagram

The organization of the vulnerabilities might suggest system upgrades that address multiple features. The ASDs should be reviewed to ensure that all of the vulnerabilities for the specific path elements are addressed and that the upgrade package ensures that all paths of the ASD are protected at least as well as in the scenario that was considered to be the most vulnerable (used for the system effectiveness analysis). Finally, each upgrade package should be evaluated with a system effectiveness analysis to ensure and measure actual improvement in protection system effectiveness. If the security system upgrade packages cannot ensure a high enough level of effectiveness to prevent the undesired events, strategies for mitigating consequences should be explored.

9.4   STRATEGIES FOR MITIGATING CONSEQUENCES

The first step in consequence reduction is to carefully examine the consequence values that were estimated for the undesired events in the analysis (see Chapter 5, "Consequence Analysis") and to identify the consequence categories that provided the highest contributions to those estimates. Specifically, consequences should be summarized by those that are: (1) *people-related,* such as loss of life and casualties, (2) *building- and facilities-related,* such as loss of properties, (3) *revenue-related,* such as loss of jobs, loss of income, loss of mission and work space, (4) *community-related,* such as loss of services, loss of public space, loss of confidence, (5) *interdependency-related,* loss of continuity of government services, loss of communications, loss of utilities, loss of emergency services, and (6) *other-related,* such as collateral damage, cascading effects, loss of recreation and cultural amenities, and the like. Summarizing consequence in this way will provide guidance as to which consequence categories should be addressed to optimize consequence mitigation.

Strategies to mitigate consequences include:

- Construction hardening
- Redundancy
- Optimized recovery
- Emergency planning

9.4.1 Construction Hardening

A specific strategy for consequence mitigation is the hardening of the construction of the building or facility against blast effects.

9.4.1.1 Blast Design Basis Threat and Explosive Scenarios

The development of a blast design basis threat for a variety of explosive attack scenarios currently appears to raise significant issues about what to do and where to apply any hardening against blast effects. Numerous government organizations involved in the defense of weapon systems against air blast and accompanying ground motions have produced manuals and methodologies for dealing with the question of hardening for protective construction systems. The use of heavily reinforced concrete and steel designs is well within the state of the art for architects and engineers to apply where needed. However, the threat criterion that is applied to meet blast requirements to protect a building and any of its critical facilities is not an easily derived factor without careful threat analysis and risk acceptability decisions that ONLY the owner and stakeholders can make. If a government agency is involved, the threat criterion that is used for blast protection design purposes is usually better defined by the agency's command and intelligence gathering structure. However, for nongovernment buildings, the blast design basis threat must be defined through a thorough understanding of the results following a full risk assessment process. A blast design basis threat is established by the owner and stakeholder decision makers, with advice from the risk assessment team. The threat is defined in terms of the amount and the type of explosives and the delivery system(s) involved. Once the blast design basis threat is established, the blast protection designer/engineer can find alternative solutions using available standoff distance(s) and hardening techniques and measures that are applicable to the building or facility to provide some level of protection against blast effects.

9.4.1.2   Site Features, Orientation, and Viable Targets

The site features, orientation, and possible obstructions may affect the adversary's access or line of sight to critical assets. Site features that might be considered as hindrances to an adversary's attack, include rigid and energy-absorbing (frangible) barriers at the perimeter such as walls, high curbs, planters, bollards, trees, shrubs, ditches, soil-rock berms (such as gabion walls), rocks, massive equipment and vehicles, and natural and man-made barriers. The building elements that might require protection or hardening include building structural members (columns, walls, beams, buttresses), utilities (cables, pipelines, values, switches, control panels, pumps, hydrants) personnel (occupants, visitors), functions (administrative, computer center operations, command and control centers, other operational missions), and entrances (portals, gates, lobbies, doors, tunnels, arches, vents, conduits). Explosive attacks on building elements might cause a catastrophic failure, partial collapse, or severe damage to the building system and/or its operations, including casualties and fire. Possible attack modes include ground-mobile devices such as positioned explosives, mechanical equipment and hand tools, ramming vehicles, and air- or water-borne projectiles.

9.4.1.3 Disruption, Damage, Total Collapse

The assessment regarding mode of failure of a building and its facilities is an important part of an evaluation and greatly assists with the determination of the consequences of a blast attack. The mode of building failure is particularly critical because of the potential for loss of life and total loss of mission that could result from a collapsing building. For example, single points of failure, such as a column, beam, or shear wall that if it fails, could initiate progressive collapse of the building, are considered critical assets that warrant not only careful analysis but also adequate protection, especially against an explosive attack. The Murrah Federal Building in Oklahoma City, damaged by an explosive attack, is considered a classic example of a catastrophic failure from progressive collapse.

During an assessment of the consequences of an attack, the question must be addressed whether disruption of service, partial damage, or total collapse is the mode of failure considered most likely to occur. The obvious exposure of a key structural element, such as a column or beam, is an indication that a careful evaluation is necessary.

An example of a disruption of service is an attack against a utility system, such as the building commercial power substation, or another building facility, such as an attached parking structure or covered walkway, where very few fatalities would be expected but certainly the building operation would be disrupted until such time as necessary repairs could be completed.

9.4.1.4 Protective Systems and Potential Upgrade Techniques

There are several techniques that may provide differing levels of hardening if the concern for protection from an explosive attack is found to be a critical consideration and the decision is to proceed with some form of upgrade or retrofit. Before analysis can proceed, the amount and type of explosive and the distance to the building and facility are required. The engineering concern relates to the potential for partial, progressive, and/or total collapse where the air overpressures and shock from the blast exceeds the strength and resistance of the existing structural systems. It is thus important to perform a preliminary structural analysis to evaluate weaknesses of elements of the building before any potential hardening techniques can be ascertained.

The objective of the building hardening strategy must also be established, which might include any of the following: (1) remain standing and entirely intact except for slightly damaged façade and some window breakage, (2) remain standing while allowing for some damage to structural elements if rapid evacuation of all surviving occupants can still be ensured, (3) remain partially standing, without progressive collapse, while allowing for most of its occupants to evacuate to "safe havens" to be rescued as soon after the attack as possible. Although selecting one of these protective strategies against a blast attack is challenging for decision maker(s), several different levels of risk and uncertainty can also be assigned to the ultimately selected hardening technique that still meets the design objective. The hardening techniques that are available range from structurally strengthening exposed critical members such as columns, beams, floor slabs, and walls using fiber-reinforced polymer composites to providing for redundant structural members that protect against failure or progressive collapse. More commonly used hardening techniques are applied in the field in the form of wraps around columns and coatings on walls and floors, using the appropriate bonding procedure, resulting in a composite that indeed strengthens the weaker structural members.

9.4.1.5 Alternative Plans, Standoff Distances, and Access Control

If the blast-hardening techniques discussed above are evaluated and found to be prohibitively expensive or not cost-effective in their specific application, alternative strategies and plans might be evaluated that may well provide some degree of protection but may not include actual hardening of the structure. An assured means of keeping any large quantity of an explosive device at a distance from key structural members is certainly useful and advantageous. The chance of damage to a structural element in a building is significantly reduced (exponentially) when a standoff distance can be assured through some means (also depending on the amount of explosives used in the attack). Various barrier systems discussed above that provide an obstruction to vehicles carrying explosives provide a viable means of protection. Their effectiveness against an attack can now be best evaluated through analysis that offers some level of confidence that building destruction or severe damage as a function of distance would be extremely difficult. The primary consideration in a design to meet a criterion for a specific standoff distance is to ensure that there will be no penetration of the perimeter by an adversary's vehicle loaded with explosives. Therefore, some means for controlling vehicle access inside the perimeter is required, through a security gate, sally port vehicle inspection station, and/or driver authentication procedures. It is important that the design criteria for standoff distance and security requirements be well integrated and that both support the security strategy and design objectives.

9.4.1.6 Calculation Capabilities, Expertise, and Resources

Several levels of computer calculations can be performed to evaluate the effects of explosive attacks on buildings and facilities.

The basic parameters that are normally required to perform such calculations to determine the extent of structural failure, displacement, and/or partial damage include: (1) the characteristics, shape, energy-release efficiency, and quantity of explosive material; (2) the distance the explosive material is placed away from the target (building and facilities); and (3) the technical description of the structure under attack, including construction materials, type of building, configuration of the key structural members, and their respective dimensions and strength properties.

The most technically sophisticated level of analysis for blast-structure *coupled* interactions use primarily finite-element-based computer simulations. These computer simulations calculate the blast environment and impact on the structure, such as overpressure and its duration (impulse), that is produced by the detonation of a well-known and easily manufactured energetic material such as the mixture of ammonium nitrate and fuel oil (ANFO), or other standard explosives such as tri-nitro toluene (TNT), or C-4 plastic explosives, to mention a few. The energy released and the accompanying blast impacts from these characterized explosives (using a hydrodynamic code such as CTH) are then coupled by computer simulation techniques (using ZAPOTEC, as an example) with the finite element structural response codes of actual three-dimensional buildings and facilities (such as the PRONTO 3-D structural dynamics code). These calculations produce detailed assessments of the damage resulting within the structural members, such as the percent strain in the materials, displacements of structural components, and resulting modes of failure that represent expected material behavior response derived from the principles of fracture mechanics and material behavior properties.

The next level of computer simulations for blast effects are less sophisticated and more simplified in determining structural response from blast than those described above. Damage results are more general and qualitative. For example, the levels of damage are usually summarized into three general categories: (1) total destruction and/or building collapse with large loss of life and property, and the building is not recoverable after the event; (2) medium damage to the building that includes some loss of life, medium structural member failures, no building collapse, and the building being potentially repairable for future use after the event; and (3) minor to no damage with no loss of life, easily recoverable from, and with minor repairs to restore operations. This level of analysis determines a first-order magnitude estimate of the damage severity and the possible requirement for alternatives for protection. Included in the alternatives are: (1) increasing the standoff distance, (2) hardening to mitigate the expected higher consequences, and/or (3) revising the criterion to some lower level of explosive threat and accepting the higher level of risk for a higher level of consequences that might occur.

The third level of model simulation for blast effects, which is less sophisticated and further simplified for the user, applies the extrapolations from graphs of blast effect curves that are based on performance of ideal explosive material quantities and standoff distances from a target. The series of curves that are determined by calculating the overpressure and impulse from a series of quantities of ideal explosive charges, such as TNT, are plotted on a graphic display to indicate the rate at which the explosive effects decay as a function of distance away from the source of the detonated material. These curves can then be used to predict, in a generic way, the degree of potential damage that might occur to buildings and facilities in the event of a blast impact on structural systems and components, including concrete or steel buildings, masonry walls, glass windows, and human bodies subjected to the blast loads.

The U.S. Treasury Bureau of Alcohol, Tobacco and Firearms (ATF) has produced for general application a generic table of explosive blast effects that are expressed in terms of the distances

ATF	VEHICLE DESCRIPTION	MAXIMUM EXPLOSIVES CAPACITY	LETHAL AIR BLAST RANGE	MINIMUM EVACUATION DISTANCE	FALLING GLASS HAZARD
	COMPACT SEDAN	500 Pounds 227 Kilos *(In Truck)*	100 Feet 30 Meters	1,500 Feet 457 Meters	1,250 Feet 381 Meters
	FULL SIZE SEDAN	1,000 Pounds 455 Kilos *(In Truck)*	125 Feet 38 Meters	1,750 Feet 534 Meters	1,750 Feet 534 Meters
	PASSENGER VAN OR CARGO VAN	4,000 Pounds 1,818 Kilos	200 Feet 61 Meters	2,750 Feet 838 Meters	2,750 Feet 838 Meters
	SMALL BOX VAN *(14 FT BOX)*	10,000 Pounds 4,545 Kilos	300 Feet 91 Meters	3,750 Feet 1,143 Meters	3,750 Feet 1,143 Meters
	BOX VAN OR WATER/FUEL TRUCK	30,000 Pounds 13,636 Kilos	450 Feet 137 Meters	6,500 Feet 1,982 Meters	6,500 Feet 1,982 Meters
	SEMI-TRAILER	60,000 Pounds 27,273 Kilos	600 Feet 183 Meters	7,000 Feet 2,134 Meters	7,000 Feet 2,134 Meters

Figure 9.2   Generic Table of Explosive Blast Effects.

concerning lethal impacts, minimum distances for evacuation requirements, and falling glass hazard distances from a target under attack (see Figure 9.2). These data are effective in planning for standoff distances from an explosive attack and developing emergency action plans for minimizing blast effects on humans anywhere in the area of the attack. Other, more specific, blast effects data on structures, humans, and equipment, although available, are restricted for use by authorized personnel who have the need and expertise for its appropriate application.

A blast analysis consultant with the appropriate credentials and experience will provide suggestions on the analyses that will assist in determining the best calculation options available for the specific building and circumstances. The less sophisticated calculations might be preferred initially to scope the level of damage that is anticipated and to estimate the order of magnitude of costs that would be involved if hardening options were to be pursued. There are numerous institutes, companies, and government agencies that have clear expertise to assist with blast analyses as appropriate. Detailed review of their past analyses and interviews with their experts will benefit the project by allowing the experts to suggest options on how to proceed with a more cost-effective approach, one phase at a time. The final analysis would entail the determination of the appropriate hardening techniques that would provide the protection that is deemed most effective, if any at all.

9.4.1.7 Decision-Making Issues and Supporting Data

Finally, the analysis must conclude whether or not blast protection does in fact provide the risk reduction that meets the expectations of the owners/stakeholders and is the best alternative for the credible threat scenarios that are described. This decision depends clearly on the issues of: (1) the credible threat and the likelihood of attack, (2) the size of the explosive charge predicted to be delivered and the determination of the location of the specific critical targets, and (3) whether there is absolute consensus that there are no other options to mitigate consequences in the event of a destructive attack.

9.4.2 Redundancy

9.4.2.1 Redundancies and Backup Systems

Building and facility redundancies and backup systems are normally considered for critical assets that if interrupted by a malevolent attack would seriously jeopardize the mission of the organization. Typical examples of redundancies and backup systems include: (1) electric power using a backup generator or uninterrupted power supply (UPS), (2) alternative external sources of commercial power to the building with secondary transformer banks at local substations, (3) dual external supply lines for water sources and dual water pipeline distribution inside the building for firefighting, (4) redundant air-cooling systems for critical computers in data centers, (5) dual security and fire alarm systems for building management, (6) alternative communication systems, including telephones, radio, and cell phones, (7) backup equipment such as computers, motors, fans, transformers, that are usually located near the operational areas for rapid replacement, (8) other miscellaneous redundancies that pertain sometimes to staff, materials and supplies, shipping and receiving, ingress/egress, and the like. The key security requirement for redundancy is that the systems need to have as large as possible separation distances to ensure that a single-point attack does not destroy both the primary and redundant systems at one strike.

9.4.2.2 Inventory and Stockpile Plans

The primary security requirement for inventory and stockpile planning is the assurance that adequate equipment and supplies are readily available in the event of an attack, to accelerate the recovery operation. This capability is especially important if equipment and materials cannot be readily obtained and in sufficient time to successfully recover from an attack. Of particular interest are items that have a long lead time for delivery such as transformers, HVAC chillers, electric controls and switch gear, and other special equipment that might have only one supplier and that one in a foreign country. Stockpile plans would consider emergency materials that need to be readily available at the site or nearby, such as water for firefighting and first aid and medical supplies, including vaccines in the event of a biological attack, respirators and masks for evacuation purposes, and the like. The value of doing a check of inventory and stockpile plans is that this provides assurance that if mitigation strategies depend on this material, its availability on-site, which might be the most critical requirement to minimize losses is assured.

9.4.2.3 Material and Contract Support Plans

Related to the inventory and stockpile plans, there is specifically the need to have prearranged contracts and support agreements to provide for the needs of the building and security manager immediately upon the initiation of recovery plans following an attack. The planning that is required here refers to the identification of what activities can be accomplished with in-house resources, including additional personnel, and the activities that will require the support of others.

9.4.3   Optimized Recovery Strategies

9.4.3.1   Backup and Alternative Projects

Backup and alternative projects refer to those similar buildings and facilities that can be made available to perform the mission in the event the primary building is destroyed. This provides for a continuity of operations, provided that a sufficient number of personnel survive or others can be hired, at short notice, to perform the tasks required to reduce the impact to the mission. A good example of this type of alternative project applies to an emergency operations center (EOC) that might also be attacked and deliberately destroyed to confuse and complicate a recovery operation. An alternative EOC project can be preestablished and ensures that the necessary equipment and systems for the mission are ready for use in an emergency. An alternate mobile command center could be helpful to reduce initial costs and maintenance and operations considerations. Some agencies have multiple buildings in the region and plans for an alternative backup to allow for more rapid recovery.

9.4.3.2   Options for Project Control Centers and SCADA

Some building missions operate with a project control center concept that is the nerve center of the organization. Within this designated control center usually rests the supervisory control and data acquisition (SCADA) function, as well as other operations, communications, and controls for mission purposes. For example, an EOC for a community would obviously be a critical asset and a potential target to attack using a variety of different scenarios, including "preattack" destruction. The objective of studying a control center and the SCADA mission, contents, and systems is to evaluate the impact if it is destroyed and establish if there are alternative plans for restoring its function in a short time period, as needed. Some critical control centers that are used for distribution of vital services, such as electric power, water resources, gas and oil, and the like, are designed to have alternate sites that can begin operations at the "flick of a switch." The more common administrative buildings and facilities might have plans that rely on outside control center resources and support agreements for backup. For example, SCADA systems that are used for remote and automatic controls, such as those for dams and lock operations, water resources and power transmissions and distributions, and petroleum materials pipeline conveyance and distributions, are often equipped with manual overrides that can be used in the event of a cyber- or physical attack, once it is detected by the operators. Such SCADA systems located inside buildings or supporting facilities, even if not important for internal building management and operations, might be critically important for systems outside the building's perimeter and require investigations for needed protection or possible backup.

9.4.3.3   Customer Agreements and Clarity of Strategy

Customer agreements in this regard deal with those outside customers that are dependent on the building owner for continuous service. If these types of agreements are seriously binding, penalties and liability issues might occur in the event of an attack that is not followed by immediate restoration of service. Mitigation strategies in this situation would pertain to the urgency of recovering to meet the customer's expectations and ensuring that there are no further cascading damages incurred by the interruption of those services. The strategy to accommodate this requirement might entail a stipulation in the agreement that in the event of an interruption of service, the supplier will restore services in a prescribed number of days or weeks, perhaps depending on the mission and the extent of the damage. The importance of clarifying the mitigation and/or recovery strategy, because a service agreement is at stake, is apparent if the objective is to minimize any consequences to the building owner, stakeholders, and customers.

9.4.4   Emergency Planning

9.4.4.1   Emergency Action Plans

Emergency action plans play a vital role in consequence mitigation. Most organizations have emergency action plans to address a natural disaster or other emergencies. These plans are useful in reviewing and understanding the existing actions that are taken following an event and in identifying critical elements, such as water sprinklers, fire alarms, evacuation routes, and the like. Any weaknesses that are discovered, especially those deficiencies that were not addressed because malevolent threats were not considered previously, must be examined for mitigation. When the security operations plans are revised, emergency action plans must be reviewed and updated to address all existing mitigation strategies.

9.4.4.2   Early Warning Systems and Evacuations

Early warning systems are in operation today at locations where naturally occurring events can be monitored and their direction of impact is predictable. In the event they occur, advance warning is given to occupants that a storm or flood is jeopardizing the area and evacuation from the premises is suggested. In the event that early warning can be applied to a malevolent attack, all efforts must be made to take advantage of alerting all occupants to evacuate or assemble, for example, in a safe haven designated in that building for that purpose. In order to be effective, all early warning systems must first be linked to early detection of the event, and secondly, to the occupant notification and broadcast system.

9.4.4.3 Temporary Security Response Force and Positioning

When appropriate, the federal government alerts the nation that a heightened state of threat is recognized from credible intelligence data; many federal agencies are then prepared to take the necessary actions to protect their facilities against a potential attack. With guidance from a completed risk assessment evaluation, the critical assets within a building have been identified, and some additional security measures would most likely be planned or in place to protect the critical asset(s). If these countermeasures are not cost-effective, or if funding is not available to install a permanent security system, other options must be considered. The advantage of placing a temporary security response force at or near the critical asset(s) for the duration of the federal alert period, or for longer periods, is worth considering. It is wise to have all the planning ready for this option in advance and any required support contracts in place or readily available. Upon notification, additional response force personnel can be positioned at the sites identified for 24/7 coverage, if necessary.

9.4.4.4 Law Enforcement Tactics

It should be noted that law enforcement tactics are important, to detect, apprehend, arrest, and neutralize the attacking adversary. If support from a law enforcement agency is required to augment an existing security force at a building or to intercept the adversary during periods when there is no security force on duty, security operation plans must be established to provide specific guidance on how local law enforcement can support the needs for the building. The tactics to be applied and the equipment that would be used must be discussed with the building's security manager well in advance so that instructions are clear and notification links are well established.

9.4.4.5   First Responder and Equipment

The first responders are the most critical personnel at the scene of an emergency when it comes to saving lives, minimizing casualties, and protecting against further destruction. As first responders, these individuals place their lives at risk because of the unsafe and insecure conditions that they are frequently required to face. In order to provide a safer and more secure work environment for them in advance, serious consideration must be given to how building configurations, access points, ingress/egress routes, and prestaged equipment can assist their operations. Detailed discussions and building "walk-around" sessions and personal interviews with first responder representatives will highlight their needs and assist the security manager in preparing for mitigation strategies as a risk reduction measure. As a result of recent building disaster site experiences, police and fire chiefs, medical officers, and emergency management personnel have clearly identified appurtenances that can be installed for facilitating rescue and firefighting operations. Among the more popular features include such structures as concrete pads for larger cranes and fire equipment, easily accessible critical shut-off valves at the street level, gates, openings, stairwells, and elevators that are operable during emergencies, first-aid stations in a protected area, and safe havens where occupants can assemble and be treated, if required.

9.4.4.6   Local Support and Agreements

In spite of the arrangements that exist between local building owners and municipal fire, police, medical, and emergency management departments, local support agreements are beneficial in terms of defining the details of the support and the conditions that would better foster cooperation and coordination. Ensuring that support agreements are current and viable for any emergency at a building is certainly a positive action for mitigating consequences.

9.5 COMBINATIONS OF REDUCTION STRATEGIES

Sometimes implementation of single strategies either to improve protection system effectiveness or to mitigate consequences alone cannot reduce the security risk to an acceptable level or cannot reduce the security risk cost-effectively. Combinations of risk reduction strategies might be explored. Using combinations of both security system upgrades and consequence mitigation strategies to reduce the risks involved might be beneficial to the overall protection of the building and its mission.

An example of a risk reduction strategy that includes both PPS features and consequence mitigation features is discussed with the following example. In the event that the threat of concern is a chemical-biological-radiological (CBR) event against the building and facilities, further examination of the HVAC systems, and especially the air-intake openings and systems, is required. All points of potential entry of any contamination into the building must be protected to minimize the exposure of personnel, property, and critical equipment. Early detection of contaminants entering the building and subsequent shutdown of the ventilation systems would provide some mitigation and will most likely minimize the effects to occupants and equipment. Applying some means of internal positive pressure and filtration upon detection of a contaminant can also be effective, depending on the equipment involved and the duration of the exposure. In addition, establishing evacuation plans for personnel in the event of a CBR attack is crucial so that the safe haven chosen does not cause more exposure of the personnel, but provides first aid, fresh air, and necessary protective gear. Protecting personnel and computer equipment

PPS= Physical Protection System

against corrosive contaminants such as sulfuric acid, chlorine, ammonia, and certain hydrocarbons requires careful investigation prior to proceeding. Biological and radiological contaminants are under investigation by numerous agencies of the government, and definitive information and protection guidance have not been formulated at this time.

The possibility that implementation costs can also be reduced if both reduction strategies are applied might provides another incentive for pursuing the combination. The usefulness of the complete risk assessment becomes more apparent when alternatives and options are reviewed and estimates are made of implementation, including operations, and maintenance costs and the potential for risk reduction are also compared against the current risk level. The return on investment and capital is a critical management objective and plays a key role in making the decision on what combination of alternative measures makes the most sense, provides the largest risk reduction, and results in the best cost-effective solution. Cost-benefit analyses introduced at this stage can also assist with the determination of the best course of action to pursue.

At this point, the basic risk assessment has been completed, including the analysis of various upgrade packages. A package can be assembled for presentation to the risk managers, including: the threat analysis, consequence analysis, system effectiveness analysis for both the physical and cyber-protection systems, estimated security risk level, and identification of strategies to reduce risk.

9.6 SUMMARY

In this chapter, strategies were discussed to reduce security risk. The two primary methods for risk reduction offered include increasing the security system protection effectiveness using physical upgrades and administrative and operational security options or consequence reduction using some form of mitigation strategy. Risk reduction measures designed to reduce the likelihood of an

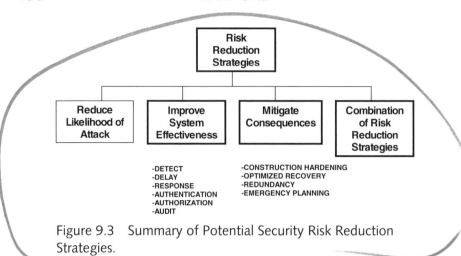

Figure 9.3 Summary of Potential Security Risk Reduction Strategies.

attack were not developed because of the unpredictable effectiveness of deterrence. Figure 9.3 summarizes possible security risk reduction strategies, that is, the protection system functions to be addressed to improve system effectiveness and strategies to mitigate consequences.

9.7 REFERENCES

1. American Society of Civil Engineers, "Structural Failures: Modes, Causes, Responsibilities," ASCE National Meeting on Structural Engineering, Cleveland, Ohio, April 1972.
2. Attaway, Stephen W., Matalucci, Rudolph V., Key Samuel W., Morrill, Kenneth B., Malvar, L. Javier, and Crawford, John E., *Enhancements to PRONTO3D to Predict Structural Response to Blast*, SAND2000-1017, Sandia National Laboratories, Albuquerque, NM, May 2000.
3. Garcia, Mary Lynn, *Design and Evaluation of Physical Protection Systems*, Butterworth-Heinemann, Burlington, MA, 2001.
4. Garcia, Mary Lynn *Vulnerability Assessment of Physical Protection Systems*, Butterworth-Heinemann, Burlington, MA, 2006.
5. ICBO Evaluation Services, Inc., "Acceptance Criteria for Concrete and Reinforced and Un-reinforced Masonry Strengthening Using Fiber-Reinforced Polymer (FRP), Composite Systems." AC125, Whittier, California 90601, January 2001.

6. Matalucci, R. V. and Miyoshi, Dennis S., O'Connor, Sharon L., *An Introduction to Architectural Surety® Education*, SAND98-2086, Sandia National Laboratories, Albuquerque, NM, September 1998.

7. Matalucci, R. V., "Architectural Surety® Tutorial," Innovative Technology for Disaster Mitigation: An Architectural Surety® Conference, Washington, DC, October 27–29, 1999.

8. Matalucci, R. V. and Miyoshi, Dennis S., "An Introduction to the Architectural Surety® Program," *Proceedings of the Conference on Architectural Surety®: Assuring the Performance of Buildings and Infrastructures*, Sandia National Laboratories, Albuquerque, NM, May 14–15, 1997.

9. Sandia National Laboratories, ZAPOTEC reference Albuquerque, NM, May 2000.

10. U.S. Army Corps of Engineers, *Anti-Terrorism Planner (AT Planner)*, Engineering Research and Development Center (ERDC), Vicksburg, Mississippi.

11. "Vehicle Borne Improvised Explosive Device (VBIED)," ATF Vehicle Bomb Table, http://www.nationalhomelandsecurityknowledgebase.com/Research/International_Articles/VBIED_Terrorist_Weapon_of_Choice.html.

9.8 EXERCISES

1. Describe the two major elements that are used in the risk reduction process, and indicate why reducing the likelihood of an attack is difficult, if in any way possible.

2. What extent of the risk assessment effort is necessary before any risk reduction technique can be determined?

3. Where does the decision maker(s) come into the risk assessment process, and what data is required before any decision can be made?

4. What are the three physical security functions that must be addressed to improve protection system effectiveness in order to reduce risk? What are the three cyber-security functions that must be addressed to improve protection system effectiveness?

5. List some of the possible consequence mitigation strategies.
   1. What are possible consequence mitigation methods for each strategy?

2. Describe why they are or are not as useful in reducing risk.

6. List some of the recovery strategies that might be useful for buildings and facilities, and describe what actions need to be taken to ensure that the described risks are reduced.

7. Why and where would a blast analysis be required, and how would the decision be made to proceed with recommended hardening against a potential attack?

8. What are the three different types of blast analysis calculations that can be performed, and where and how would each be applied?

Chapter 10

Evaluating Impacts

The first requirement of a recommended security risk reduction strategy is that it does indeed reduce security risk, ideally to the acceptable level defined by the site. The next consideration is most likely to be the cost of implementing the recommended package. Implementation of security risk reduction strategies or packages may also impact several areas of the facility, such as operations and schedules. Often the response of the public may be a consideration in the appropriateness of a particular risk reduction strategy. There may be building- or facility-specific impacts that management will consider when making risk management decisions. These impacts can be anticipated, and evaluations of the relative severity of the impacts are included in the process because these impacts are significant considerations for risk management decision makers. Impacts are evaluated as *High, Medium*, or *Low*, with these categories defined by the site.

10.1 RISK LEVEL

Risk can be reduced by decreasing one or more of the three factors in the risk equation, namely likelihood of attack, protection system ineffectiveness, or consequences. Measuring the amount of reduction for likelihood of attack afforded by a risk reduction strategy is difficult because human behavior is notoriously variable and thus hard to predict with confidence. Risk reduction strategies most often focus on upgrading the protection system (to reduce system ineffectiveness) or reducing the consequences of a successful attack. Reductions in the likelihood of attack may well be associated with these strategies, but credit for any reduction in likelihood of attack cannot be taken because it cannot be mathematically demonstrated.

The obvious question is whether or not the risk reduction package will lower the risk value, and if so, by how much. The effect of a risk reduction strategy or package on the risk level of the building or facility is estimated by using the relevant risk assessment steps – threat analysis, consequence analysis, and/or system effectiveness assessment – to estimate the new security risk if the strategy or package were implemented.

ASDs should be reviewed to ensure that proposed protection system upgrades affect all paths. It is important that the most-vulnerable adversary path for the upgraded system be adequately protected. For example, if during the upgrade process, the protection objective includes detecting the intruder at the property boundary, every penetration of the boundary must have a means of detection. Likewise, all paths should have adequate delay. Sometimes placing delay features at the critical asset allow all paths to be affected. Reviewing the ASDs helps prevent overlooking the protection of any penetrations.

The protection system effectiveness for the upgraded system (intended to reduce system ineffectiveness) must be estimated. The same process used in Chapter 7, "System Effectiveness," to estimate the effectiveness of the existing or baseline protection system and to identify any site-specific vulnerabilities is now used to estimate the effectiveness of the upgraded protection system in reducing or eliminating the identified vulnerabilities.

Path Elements for Most Vulnerable Scenario	Detection Level		Delay Time (seconds)	
	Baseline	**Upgrade**	**Baseline**	**Upgrade**
Gate (pedestrian)	*Low*		*0*	
PropertyArea			*7*	
Door (pedestrian)	*Medium*		*30*	
Building Interior Area			*3.5*	
Door (pedestrian)	*Medium*		*10*	
Control Room Area			*0*	
Task at Target	*Medium*		*15*	
Delay Time After Detection			A: *28.5 sec* A: *28.5 sec*	
Response Time			B: *300 sec* B: *300 sec*	
Estimated Baseline System Effectiveness Level Estimated Upgrade System Effectiveness Level A<B, System Effectiveness = Low A~B, System Effectiveness = Medium A>B, System Effectiveness = High			Baseline System Effectiveness: *Low* Upgrade System Effectiveness:	

Figure 10.1   Example Summary Form for Comparing Estimated Protection System Effectiveness (*PE*) Values for Upgrade Package to Baseline Protection System Effectiveness Values.

Figure 10.1 shows a form that can be used to summarize the estimation of P_E for the upgraded system. Also, consequence values associated with each undesired event, as developed in Chapter 5, "Consequence Analysis," should be reviewed to determine the effects of the consequence reduction features. Figure 10.2 shows a form that can be completed to compare the example baseline consequences for undesired events to the consequences for the same hypothetical undesired events occurring with the recommended upgrade package implemented. Use the site-specific Reference Table of Consequences and relative consequence values for undesired events developed in Chapter 5, "Consequence Analysis," to estimate the consequence values for each recommended package.

Figure 10.3 shows a worksheet that can be used to record the likelihood of attack, system ineffectiveness, consequence, and estimated risk values for each risk reduction package by adversary type and undesired event. Such detailed records provide an excellent paper trail for quality control purposes as well as documenting the risk assessment.

| Undesired Event | Measure of Consequence | | | | Consequence Severity | | | |
| | | Value | | By Type H/M/L | | By Event H/M/L | |
	Type	Base-line	Up-grade	Base-line	Up-grade	Base-line	Up-grade
Disruption of Operations (sabotage of vital equipment by cyber-attack)	Economic loss (property loss + revenue)	$3M		M			
	Economic loss (users)	0		L			
	Deaths	0		L			
	Geographic Impact	Local		L			
	Public confidence	6 mo		H			
				Enter highest conse-quence		H	
Theft of Valuable Asset(s) precious metals)	Economic loss (property loss + revenue)	$3M		M			
	Economic loss (users)	0		L			
	Deaths	0		L			
	Geographic Impact	Local		L			
	Public confidence	1 day		L			
				Enter highest conse-quence		M	
Crimes against People (hostage situation)	Economic loss (property loss + revenue)	0		L			
	Economic loss (users)	0		L			
	Deaths	0-1		L			
	Geographic Impact	Local		L			
	Public confidence	None					
				Enter highest consequence		L	

Figure 10.2   Example Summary Form for Comparing Estimated Consequence (C) Values for Upgrade Package to Baseline Consequence Values.

Finally, the summary form provided in Figure 10.4 compares the baseline (or original) system risk to the upgraded system risk for the threat groups that apply. This summary form should be prepared for each recommended risk reduction strategy.

Undesired Event	Measure of Consequence			Consequence Severity			
	Type	Value		By Type H/M/L		By Event H/M/L	
		Base-line	Up-grade	Base-line	Up-grade	Base-line	Up-grade
Destruction of Building (vehicle bomb)	Economic loss (property loss + revenue)	$7M		H			
	Economic loss (users)	0		L			
	Deaths	10-20		H			
	Geographic Impact	Local		L			
	Public confidence	6 mo.		H			
				Enter highest consequence		H	

Figure 10.2   (continued)

P2122 (handwritten)

RISK CALCULATION WORKSHEET (RISK REDUCTION PACKAGE)

Likelyhood on an Attack (handwritten, top)   *Ineffective* (handwritten, right)

ADVERSARY TYPE _____

| Date: | Recorded by: |
| Facility Identifier: | Package Identifier: *Consequence → System* (handwritten) |

Undesired Event	P_A	C	$1 - P_E$	RISK
Disruption of Operations (sabotage of vital equipment by cyber-attack)				
Theft of Valuable Asset(s) (precious metals)				
Crimes against People (hostage situation				
Destruction of Building (vehicle bomb)				

Figure 10.3   Example Risk Calculation Worksheet for Risk Reduction Packages.

10.2   COSTS

Probably the single most important impact (after reducing security risk) of implementing security risk reduction packages is the cost. While a detailed and precise cost estimate would be premature, costs for packages will be compared by management decision makers. The dollar cost of packages will be assigned a value of *High, Medium,* or *Low,* with each category representing

SYSTEM RISK COMPARISON (Original vs. Upgraded) Summary											
Date:	Recorded by:										
Facility Identifier:	Package Identifier:										
	RISK VALUES, by Adversary										
	Original System → ◻▨ ← Upgraded System										
Adversary → / ↓ Undesired Event	Terrorist										
	International	Domestic	Militia/Paramilitary	Extremists	Criminals	Gangs	Vandals	Insiders	Other		
1. Disruption of Operations (sabotage of vital equipment by cyber-attack)											
2. Theft of Valuable Asset(s) (precious metals)											
3. Crimes against People (hostage situation)											
4. Destruction of Building (vehicle bomb)											
5.											
6.											

Figure 10.4   Example System Risk Comparison Summary Form.

a range authorized by the risk manager or other management stakeholder. After costs are estimated for each proposed upgrade package, a means to record the costs for easy comparisons is useful. Figure 10.5 shows graphs that can be used to present costs for several packages. (As evident from figure, this format can be used for several impact areas, not just cost information.)

The bar graph in Figure 10.5 can be shaded (manually or electronically) with appropriate values to allow comparison of relative costs associated with upgrade packages: L = *Low*, M = *Medium*,

IMPACTS OF SECURITY RISK REDUCTION PACKAGES			
Date		Recorded by:	
Facility Identifier:			
Impact Level *Circle level and shade graph below*			
	Package 1	Package 2	Package 3	Package 4

	Package 1	Package 2	Package 3	Package 4
Cost	L M H	L M H	L M H	L M H
Operations/ Schedules	L M H	L M H	L M H	L M H
Public Opinion	L M H	L M H	L M H	L M H

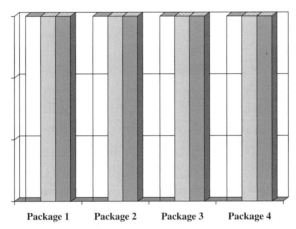

Package 1   Package 2   Package 3   Package 4

Figure 10.5   Displaying Relative Impacts of Multiple Risk Reduction Packages.

and $H = High$. Individual sites define the cost values for L, M, and H, thus allowing the decision makers to establish cost boundaries for the recommended risk reduction packages within the available resources.

10.3   OPERATIONS/SCHEDULES

Implementation of an upgrade package could have a negative impact on operations or schedules if it imposes delays or significant changes in normal practices. An estimate must be made of the impact on operations imposed by the risk reduction package(s). The form shown in Figure 10.5 for recording costs can also be used to display the relative impacts to operations and/or schedules of the recommended risk reduction packages. For the example form shown in Figure 10.5, the middle bar can be shaded manually or electronically to compare disruptions to operations and/or schedule for the protection system upgrade package(s) and consequence mitigation packages recommended to reduce the estimated security risk. On this form, L = *Low*, M = *Medium*, and H = *High*; individual sites define the impact values for L, M, and H in order to render this comparison more valuable for decision support.

10.4   PUBLIC OPINION

Public opinion or political relations can be sensitive to some risk reduction packages. Credibility and acceptance by the public are important. An estimation of the impact on public opinion imposed by the upgrade package(s) must be made. Figure 10.5 can be used to compare the impacts of each risk reduction strategy on public opinion, using the relative terms L, M, and H. Decision makers or risk managers at the building or facility of concern define the impact values for L, M, and H.

10.5   OTHER SITE-SPECIFIC CONCERNS

Risk reduction packages could cause concerns that are site-specific. Some of these could be impacts on facility reliability, rate payer vs. the taxpayer issues, political sensitivities, environmental concerns, and the like. The risk analysis should identify any other sensitive issues that could result from upgrading the protection system or reducing the consequences of a malevolent attack. The graphs in Figure 10.5 could be labeled and completed for any site-specific concerns.

10.6   REVIEW THREAT ANALYSIS

To this point, the project-specific threat has been used for the analysis. After reviewing the comparative risk values for the upgraded system and the baseline system and the results of the impact analysis of the risk reduction packages, a subset of the project threat may be selected for the design threat. It may not be possible to reduce the risk to an acceptable level by upgrading the protection system or reducing the consequences. The impact of cost or any of the other parameters may require that a lesser threat be addressed. This revised threat spectrum is a decision that specifies what will be protected against at this time. For example, if a risk reduction package is selected for implementation, the decision may be made to establish the design threat as the adversarial groups that are addressed by that particular upgrade package. The remainder of the threat spectrum would then be addressed permanently on a future schedule and addressed immediately by contingency security measures.

Complete Figure 10.6 to summarize the revised threat if it is different from the project-specific threat.

Revised Threat Description					
Date:			Recorded by:		
Facility Identifier:					
Type of Adversary	Number	Equipment	Vehicles	Weapons	Tactics

Figure 10.6   Revised Threat Description Form.

10.7 SUMMARY

The risk reduction strategies and packages recommended to reduce the security risk at the subject building and facility do not exist in a vacuum. Implementing an upgrade to the protection system to prevent a successful adversarial attack or mitigating the consequences of a successful attack will have impacts at the facility or building beyond the security realm. These impacts must be analyzed as they affect risk management decisions. The risk level before and after implementation should be reviewed to ensure that the risk reduction strategy or strategies to be implemented do, in fact, reduce risk to an acceptable level. Other impacts to be analyzed include costs, operations and schedules, public opinion, and other impacts that are specific to the site. This chapter provides tools for analyzing those impacts.

In the event that the impact analysis shows that risk reduction strategies implemented at the building or facility are not adequate to reduce the security risk to the acceptable level for the site-specific threat, a subset of the threat may be selected to be addressed at the current time. The balance of the site-specific request would be addressed as soon as possible.

10.8 REFERENCES

1. Biringer, Betty, "Risk Assessment Method for Electric Power Transmission," presented at Carnahan Conference on Security Technology, sponsored by IEEE, Albuquerque, NM, October 2004.
2. *Sandia National Laboratories Security Risk Assessment Methodologies*, http://www.sandia.gov/ram.
3. Matalucci, Rudy and Strothman, John, "Security Risk Assessment Procedures: Countering Terrorism and Other Threats," Infrastructure Security Course sponsored by ASCE and Sandia National Laboratories, Las Vegas, NV, January 26–27, 2006.

10.9 EXERCISES

1. Why is it necessary to consider impacts that are not related to security issues when considering security risk reduction schemes?
2. List several examples of possible site-specific impacts of security risk reduction schemes.
3. If security risk cannot be reduced to an acceptable level by upgrading the protection system, what other alternatives are there?

Chapter 11

Risk Management Decisions

This risk assessment process was developed to support risk management decision makers. At this point in the Security Risk Assessment and Management Process, two steps remain: generation of a report of risk assessment results for presentation to management and risk management decisions.

11.1   INTRODUCTION

The security risk assessment effort culminates in the presentation of the results to the decision makers or other requesting management representatives. The decision makers are provided with the description of the threat, an estimate of the current security risk, recommendations for reducing that risk, and an analysis of the impacts of implementing each potential risk reduction package, including an estimation of the reduction in risk afforded by each package. All supporting documentation is provided to the decision makers.

The site-specific estimated security risk to the subject facility is an accurate and thorough snapshot in time. If the security risk is deemed to be too high for any given undesired event or threat, the guidance on reducing risk provides information on the

165

effectiveness of various risk reduction strategies and packages and the nonsecurity impacts of implementing each strategy or package.

The purpose of the security risk assessment is to provide the information necessary to support risk management. The decision maker reviews this material, perhaps asking for clarification or further detail, and determines whether the identified estimated security risk at the building or facility is within an acceptable limit or requires reduction.

11.2   RISK ASSESSMENT RESULTS

To support the decision makers with adequate information, the assessment team presents the estimation of security risk at the facility or building, recommended risk reduction strategies and packages, all supporting documentation, and all assumptions that were made in order to complete the risk analysis. This material is usually presented in a briefing, which affords decision makers the opportunity to request clarification or additional data and ask any questions about the assessment, including its assumptions, conclusions, recommendations, and supporting data.

In addition to the briefing presentation, a final report summarizes the security risk assessment performed at the facility. The results of the assessment and any proposed mitigation efforts are presented in the report, which contains sensitive information that must be protected. This documentation provides information that can be used by management to support resource allocation decisions. The final report is a snapshot of the comparative risk values for a specific facility from specific threats at a specific point in time. The format for the final report consists of the following sections:

- Executive Summary
- Introduction
- Threat Analysis
- Consequence Analysis
- System Effectiveness Assessment

- Risk Estimation
- Risk Reduction Strategies and Packages
- Impact Analysis
- Supporting Documentation

A brief description of the contents of each of these sections is provided below. The results discussed in this report support the recommendations, and the underlying effort that provided the summary information is available in the completed site-specific work products.

11.2.1   Executive Summary

This one- or two-page summary provides a very brief overview of the subject site-specific risk assessment and management process. At a minimum, the work that identified the likelihood of attack by an adversary, the consequences of such an attack, and the effectiveness of the existing security system in preventing such an attack should be summarized in the Executive Summary. A summary description of the recommended risk reduction packages and the associated impacts should be included. Be sure to include summary significant findings in this section, as it is likely to be the most widely read section of the report.

11.2.2   Introduction

This is a very short section that explains:

- The rationale for performing security assessments on facilities (increasing incidents, Presidential Decision Directives, management decisions, collective concern of regulating bodies, etc.).
- The sensitivity of the data included in the report and the requirement to protect the information. Specify that the fault tree and any other extremely sensitive material omitted from the final report can be viewed on a need-to-know basis.

- Why this facility was selected (high-profile facilities, threats have been received, incidents of vandalism or crime have increased, formal or informal screening by management, demonstration purposes, etc.).
- A very short description of the facility, focusing on any security issues (incidents, concerns, upgrades, failures, etc.).
- An introduction to the risk equation and its components.
- The structure of the report (Executive Summary, Introduction, Threat Analysis, Consequence Analysis, System Effectiveness Assessment, Security Risk Estimation, Security Risk Reduction Strategies and Packages, Impact Analysis, and Supporting Documentation).

11.2.3   Threat Analysis

Likelihood of attack is the first component of the risk equation. Two aspects of likelihood of attack are discussed in the final report: the site-specific threat and a summary of the threat analysis.

List the various adversaries identified as potential threats to the facility. Work products specifying further information on the number of people in the adversary group, the types of equipment available to them, any vehicles or weapons the adversary might have, and the tactics to be expected from the adversary should be included in the supporting documentations for each adversary identified at this particular facility. Identify the most significant of the undesired events and include all work products in the supporting documentation.

The reader is referred to supporting documentation for more information.

11.2.4   Consequence Analysis

Consequence is the second component estimated for the risk equation. The final report discusses the values (*High, Medium,* or *Low*) assigned to the consequence of an undesired event occurring. The undesired events are plotted against the likelihood of attack by each of the adversaries threatening the facility. The most significant of these undesired events are identified.

The reader is referred to the supporting documentation for more information and the complete summary.

11.2.5   System Effectiveness Assessment

Security system effectiveness is the third and last component estimated for the risk equation. Five aspects of the protection system effectiveness are discussed in the final report: high-priority undesired events; a site-specific ASD that models the physical protection system at the facility by identifying the adjacent physical areas between offsite and each critical asset, including any system features, such as a door or a wall, between adjacent areas; selected adversary scenarios; a cyber diagram that models the cyber-protection system at the facility; a site-specific vulnerabilities summary; and a table that summarizes the security system effectiveness at the facility.

The final report should include a narrative description of any significantly easy, likely, or damaging adversary scenario. It should identify and include a narrative description of any significantly easy, likely, or damaging adversary scenarios/undesired event/ system vulnerability combinations.

The reader is referred to supporting documentation for more information.

11.2.6   Risk Estimation

The three variables of the risk equation – likelihood of attack, consequence, and protection system ineffectiveness – have been defined, described, and assigned values so that a relative security risk value can be calculated for each identified medium- or high-priority undesired event. A narrative discussion about the relative risks and whether such levels of risk are acceptable to management should be included in the final report. (Threshold risk levels are usually specified at the onset of the assessment.)

The reader is referred to the supporting documentation for more information.

11.2.7   Risk Reduction Strategies and Packages

If the security risk is deemed to be unacceptably high for any given undesired event or threat, the efforts undertaken to identify ways to reduce the risk should be described.

Upgrades to increase security system effectiveness for each undesired event must take into consideration the PPS functions of detection, delay, and response and the cyber-protection functions of authentication, authorization, and audit; ways to increase these capabilities are suggested. Upgrades to decrease the consequence of an undesired event occurring are developed if it proves impossible to prevent an adversary from causing an undesired event. The final report should include a narrative description of an adversary/undesired event combination with a significant reduction in risk associated with an upgrade package.

The upgrade package or packages that provide the most significant reduction in risk with the least negative impact are recommended. Both short-term and long-term upgrades may be recommended. These recommendations are intended to provide decision-support information for management, so it is important to justify the recommendations, based on collected and analyzed data.

The reader is referred to the supporting documentation for more information.

11.2.8   Impact Analysis

The final report provides a summary comparison among upgrade packages under consideration. Relative values for costs, impacts on operations and schedules, and public opinion are assigned so that the impact of various upgrade packages can be compared. The final report should include a narrative description of the summary comparison of the impacts of potential upgrades.

If the impacts appear to be limiting, a threat level description against which the upgraded protection system or consequence reduction package will be effective may be developed. It may not be feasible to protect against the site-specific threat, but some subset of that threat can be reflected in this threat level description.

The reader is referred to the supporting documentation for more information.

11.2.9 Supporting Documentation

All work products and supporting documentation should be organized and presented in appendices, with the exception of fault trees and other sensitive information, which should be protected. This sensitive material may be viewed upon request by decision makers with a demonstrated need to know.

11.2.10 Report Overview

This report provides the data required to support security management decisions. In addition, the report and its supporting documentation provide a way to trace accountability, a baseline record, and site-specific data with potential application to other problems or issues beyond security. Should the threat, the mission or consequences, or the security system change at the facility, the report provides information that will greatly reduce the level of effort required for subsequent assessments. The baseline record will also be very helpful for such other issues as changed requirements, resources, emphases, and/or management.

11.3 RISK MANAGEMENT DECISIONS

Risk programs use a combination of risk financing and risk control tools to manage the risk. Risk financing is primarily insurance. Risk control includes:

- Risk avoidance, which is accomplished by eliminating the source of the risk. Moving hazardous material out of a building that cannot adequately protect it to a building specifically designed to protect hazardous material is an example of risk avoidance.

- Risk reduction, which is achieved by taking action to lower risk to the building or facility to prevent or reduce the severity of the loss. This is the goal of many security programs – to lower risk by implementing security measures or, if the attack cannot be prevented, to mitigate the consequences of the attack.
- Risk spreading, which is accomplished by having similar services/processes/assets at more than one facility site. By separating assets, fewer assets are placed at risk during any given adversary attack.
- Risk transfer, which is the use of insurance to cover the replacement or other costs incurred as a result of the loss.
- Risk acceptance, which is the recognition that there will always be some residual risk and, in some cases, it nay be more cost-effective to live with the risk than to reduce it.

Deciding the appropriate response to an identified risk is the bailiwick of risk managers. The key to a successful decision is knowingly determining a risk level that is acceptable, rather than unwitting acceptance of an existing amorphous risk. The purpose of the risk assessment is to provide the decision makers with the information they need to make and support good decisions.

Informed by the risk assessment data, the risk manager can better choose whether to:

- Accept the risk. A risk manager might select this option when the consequences of an attack or undesired event are less costly in some way than preventing the attack or mitigating the result.
- Buy more insurance. If the consequences are less than devastating, this could be a cheaper way to manage risk.

- Request further analysis. Different assumptions or information may yield a more informative or useful analysis.
- Reduce risk. Risk can be reduced by increasing protection system effectiveness or by mitigating consequences. Consequence mitigation usually involves people, procedures, policies, training, and equipment. Consequence mitigation is an appealing choice for a building or facility because generally it is a more cost-effective approach for reducing risks than buying physical protection technologies.
- Establish a threat-level description that describes a subset of the site-specific threat that can be protected against right now, with plans for addressing the higher-level threats as resources permit.
- Develop a contingency protection system upgrade that can address a low-level threat all of the time and can be ramped up to address a higher-level threat when an elevation in the threat level occurs, such as an alert or emergency situation.

The risk manager uses the information provided in the briefing and final report to determine the appropriate response to the security risk. This information is intended to provide them with the data necessary to make difficult decisions concerning resource allocations for managing security risk at the subject facilities or buildings

11.4 ESTABLISH DESIGN THREAT

One of the most important products of risk management decisions is to set the level of threat for which the security system upgrade will be designed. Historically this particular threat description is called the (design threat). The design threat may be the threat description used in the risk assessment or may be some modification of it.

11.5 SUMMARY

The action items in Risk Acceptance or Mitigation belong to the decision makers. The role of the assessment team is to provide accurate security risk estimates and risk reduction strategies, supporting documentation, assumptions used in the assessment, and any other information required or requested to inform management decisions on the level of security risk that is acceptable at the particular site, building, or facility assessed and how best to achieve that acceptable level.

11.6 REFERENCES

1. Biringer, "Betty, Risk Assessment Method for Electric Power Transmission," presented at Carnahan Conference on Security Technology, sponsored by IEEE, Albuquerque, NM, October 2004.
2. *Sandia National Laboratories Security Risk Assessment Methodologies*, http://www.sandia.gov/ram.
3. Matalucci, Rudy and Strothman, John, "Security Risk Assessment Procedures: Countering Terrorism and Other Threats," Infrastructure Security Course sponsored by ASCE and Sandia National Laboratories, Las Vegas, NV, January 26–27, 2006.

11.7 EXERCISES

1. What do you think is the most important factor driving risk management decisions?
2. Do you think that insurance is a reasonable approach to managing risk against an international terrorist adversary? What about a vandal? Where would you draw the line? What information would you need to make this decision?
3. What other uses can you think of for the information in the final report? How often do you think risk assessments should be performed? What occurrences might trigger the need to update the security risk assessment?

Chapter 12

Summary

This textbook has demonstrated an analytic process to qualitatively assess security risk.

Application of the analytic process helps managers understand and manage the security risk for their facility, business, or industry. An overview of the Security Risk Assessment and Management Process is presented in this chapter. Figure 12.1 describes the process.

First a risk assessment team, made up of subject matter technical experts, is established. After the entire risk assessment process has been completed by the team, management will receive a risk assessment package which will consist of a statement of the threat description, detailed analysis of the security risks, several options for reducing the risks to acceptable levels, and an impact evaluation on total costs, operations, schedules, and acceptability.

The *risk assessment* process has ten required steps and two optional steps:

Optional – Screening analysis to prioritize corporate assets

1. *Characterize* the facility.
2. Analyze the *threat* and estimate the *likelihood of attack*.

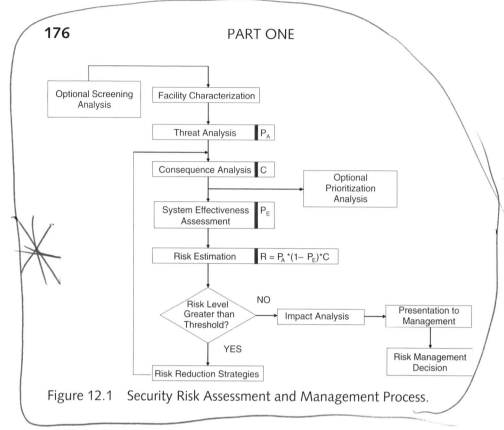

Figure 12.1   Security Risk Assessment and Management Process.

3. Estimate *consequences* from the attack.

Optional – Prioritization analysis to prioritize specific facility assets

4. Assess the *effectiveness* of the physical and cyber-protection systems.
5. Estimate *relative security risk* as a function of:
   a. Likelihood of attack
   b. Security system ineffectiveness
   c. Consequence
6. Compare estimated risk level to threshold.
7. Suggest risk reduction strategies, if estimated risk level is above threshold, followed by re-evaluating the consequences and protection system effectiveness to measure and ensure *relative security risk reduction*.

8. Analyze impacts of risk reduction packages.
9. Present results to management.
10. Risk management decisions are made.

The *risk assessment* process begins with basic facts and assumptions and each step builds on the previous step. The final results are defendable because they are traceable to the original facts, and assumptions have been documented. Results are repeatable and updates to any step are easily addressed without starting over. The *risk assessment* process can be adapted to assess the security risk for most entities. The security of dams, energy infrastructures, chemical facilities, buildings, and communities has been enhanced by the application of the Risk Assessment and Management Process. A summary of each step and its methodology follows:

Optional – Screening Analysis

A method is provided to aid owners with many facilities requiring security analysis. The reference parameter for the screening analysis is consequence level. First, a common set of undesired events is established to be compared for all of the facilities. Next, a rough estimate of consequence level is made for each undesired event for each facility. Then facilities are ordered based on the number of occurrences of the highest level of consequences. Owners can use this method to decide whether or not all of their facilities need to be reviewed immediately or to prioritize their facilities and analyze their most critical buildings first if time and resources are limited.

12.1   FACILITY CHARACTERIZATION

The products of an effective facility characterization are

- Complete facility description
- List of undesired events

- List of critical assets to be protected to prevent undesired events
- List of protection objectives for the security system

To develop these products, the risk assessment team begins their analysis of each facility by collecting and evaluating the following information: physical details; cyber-information-system details, facility operations, existing security protection systems, workforce description, and restrictions, requirements, limitations (generally regarding codes, compliance, and law). Any security level reduction measure, used to thwart attack of the critical assets, will usually include as its primary objectives, prevention of the following undesired events:

- Disrupting normal work operations
- Compromising the structural integrity of the building
- Compromising the health and safety of occupants
- Disabling or misusing the utilities
- Disabling or misusing the HVAC
- Disabling or misusing the emergency systems

With detailed information about the facility and identification of the protection objectives, a site-specific logic diagram, or fault tree, can be created and used to determine all the ways undesired events may occur at a particular facility and the critical assets to be protected. Once developed, the fault tree will represent the components and subsystems of events that can result in a specified undesired event. Identification of the assets critical to the operations' components and subsystems will logically reveal which assets must be protected in order to prevent the undesired event.

12.2 THREAT ANALYSIS

The threat analysis is usually completed by a threat specialist who has established ties and maintains contact with local, state, and federal law enforcement agencies, such as the Federal Bureau of Investigation (FBI). The five steps in the threat analysis process are:

- Collect information on the potential threat
- Derive an adversary spectrum for a given building, vicinity, or industry
- Describe adversary capabilities
- Estimate the threat potential for attack for specific adversarial groups for a given asset
- Define the adversarial threat for a given entity

In general, the adversary spectrum consists of outsiders and a single insider, who can be international terrorists, domestic terrorists, criminals, extremists, vandals, foreign intelligence personnel, psychotics, and anyone with knowledge of operations or security systems and who has unescorted access to facilities or security interests. Such information as motivation, tactics, intelligence-gathering means, targets of interest, expected number in group, equipment, transportation, weapons, explosives, technical skills/knowledge, financial resources, and potential for collusion with an insider is collected by the threat analysis and used to estimate the likelihood or potential of attack. The insider threat is defined in terms of job position and the privileges, knowledge, and access to assets and the security system afforded by the position that could be exploited.

A qualitative relative methodology can be applied to assess *threat potential*. Three factors must be considered in this analysis: adversary capability, adversary history/intent, and relative attractiveness of asset to adversary. Once this information has been collected and interpreted, numerical scores can be applied and all the scores summed and partitioned into the likelihood of attack ranges of *Low, Medium, High*, and *Very High*. Results of the threat analysis are a definition of the threat spectrum and the values used to estimate the first parameter of the security risk equation: Likelihood of Attack, P_A.

12.3 CONSEQUENCE ANALYSIS

Consequence Analysis estimates the value of a particular consequence for each undesired event for a given facility. From such measurable criteria as deaths, economic impacts, loss of assets, environmental damage, etc., it becomes possible to develop a *Reference Table of Consequences* as an objective tool. As many specific quantitative elements, such as numbers of people, dollar amounts, and the like, must be supplied for each applicable consequence. Military and industry standards can be used to supply early numbers. From the type of consequence and its extent of impact, the *severity of the consequence,* that is, *High, Medium*, or *Low,* can be deduced. The value estimates the second parameter of the security risk equation: Consequence, C.

Optional – Prioritization Analysis

Likelihood of attack and consequence level can be used for building owners to identify the assets in their building that have a high likelihood of being attacked and represent a high level of consequences if lost. A *prioritization matrix* for the site can be constructed by plotting the ordered pairs of threat likelihood of attack against the consequences of successful attack for the most critical adversary group(s). The matrix can be used to help management prioritize either a number of different undesired events for a given facility or prioritize critical assets of the facility.

12.4 SYSTEM EFFECTIVENESS ASSESSMENT

The objective of system effectiveness assessment is to estimate the effectiveness of the protection system, both physical and cyber-systems, to meet the protection objectives specified during *facility characterization*. Effectiveness assessment begins with postulated adversarial strategies to accomplish the undesired events that can be derived from the site-specific fault tree. Using the strategies, ASDs, site information, and the expert opinion of the analysis team, adversary scenarios can be suggested that are optimum from the adversary's point of view because they take advantage of system weaknesses and paths that are the least protected and easiest to accomplish. These adversary optimum scenarios are assumed to be the most vulnerable and are used to estimate the effectiveness of the protection system.

Some undesired events may be accomplished by a physical attack, some by a cyber-attack, and others by either type of attack. A system effectiveness analysis is completed for all attack modes that are applicable. For the physical protection system effectiveness assessment, the effectiveness and integration of the detection, delay, and response functions are used. Simple protection systems can be evaluated simply by checking for missing or weak features for any one of the three physical protection functions. For complex physical protection systems, the feature detection and delay times in Appendix C, "System Effectiveness Worksheets" are used to estimate system delay time after reliable detection to estimate whether or not the adversary could be interrupted before the undesired event is accomplished. If the adversary time is less than the response time, system effectiveness is judged to be low.

For cyber-protection system assessment, the effectiveness of the system features for the authentication, authorization, and audit function are estimated for cyber-paths to the critical assets.

Insider threat is more difficult to prevent and requires an investment in personnel screening, physical protection, cyber-protection, and operations security.

Finally, whenever protection system effectiveness is judged to be low, site-specific vulnerabilities are identified. The list of site-specific vulnerabilities is valuable later for suggesting risk reduction strategies if the security risk is deemed to be too high. The third parameter for estimating security risk is system ineffectiveness, $1 - P_E$, the complement of system effectiveness.

12.5   RISK ESTIMATION

Estimating relative security risk requires combining the three security risk parameters (likelihood of adversary attack, system ineffectiveness, and the consequences of the adversary's success). The traditional security risk equation (below) provides a reference level of risk (*Very High, High, Medium, Low* and *Very Low*) that is estimated for comparison purposes. Security risk is difficult to quantify because the three parameters of the equation are not random variables, and they are not independent variables, thus the mathematical rules do not allow values to be multiplied.

$$\text{Risk} = P_A * (1 - P_E) * C$$

$$P_A = \text{Likelihood of attack}$$

$$P_E = \text{System effectiveness}$$

$$(1 - P_E) = \text{System in-effectiveness}$$

$$C = \text{Consequence}$$

Associating numbers with the levels and deriving a quantitative value for the risk level is a temptation that should be avoided. Instead logic and expert judgment should be applied to estimate security risk levels based on the assessment-derived values for likelihood of attack, system ineffectiveness, and consequences. The estimated security risk value provides risk managers with a valuable baseline *measure* of security risks that can be used to make informed risk management decisions.

12.6 COMPARISON OF ESTIMATED RISK LEVEL TO THRESHOLD

Estimated risk levels are compared to a predetermined risk threshold to decide whether further analysis is required. The threshold is determined by the analysis team and security risk managers; the threshold level is a strict upper bound on the security risk level that would be considered acceptable to stakeholders.

12.7 RISK REDUCTION STRATEGIES

If security risk estimates are greater than the threshold, risk reduction strategies are explored. Risk reduction strategies focus on two of the three parameters of the security risk equation, namely system ineffectiveness and consequence. Likelihood of attack is not included because of the unpredictability of deterrence in measuring and predicting the duration of its effectiveness.

Reducing system ineffectiveness means increasing protection system effectiveness. The easiest way to increase system effectiveness is to review, summarize, and organize the list of site-specific vulnerabilities identified in *system effectiveness assessment* and then select protection feature upgrades to remove or secure the vulnerabilities. ASDs are reviewed to ensure that all vulnerable paths are covered by the upgrades. Both the physical and cyber-protection systems may be involved based on the nature of the vulnerabilities. Upgrade features will provide protection for the detection, delay, and response functions for physical protection and protection for the authentication, authorization, and audit functions for the cyber-protection system.

Upgrade features may be grouped in to packages dependent on threat level and specific concerns and conditions of the building or corporation. Usually three or more different upgrade packages are developed. Each upgrade package is then analyzed for system effectiveness. If system effectiveness is estimated to be *High*, especially, the upgraded system would be expected to *prevent* the undesired events; an *impact analysis* is completed for the upgrade package(s). If the system effectiveness is not high enough to prevent the undesired event, consequence mitigation strategies are postulated. Consequence mitigation strategies might include one or more of the following:

- Construction hardening
- Redundancy/backup
- Optimized recovery
- Emergency planning

Consequence mitigation elements can be grouped into upgrade packages and consequence analysis repeated to ensure the consequence level has been reduced. A new security risk level can be estimated for the upgrade packages comprised of improved system effectiveness features and consequence mitigation elements. System effectiveness assessment, consequence analysis, and risk estimation steps are then repeated to ensure security risk reduction.

The estimated risk level is compared to the threshold; if the estimated risk is below the threshold, an *impact analysis* is completed. If the estimated risk level is above the threshold, additional features are suggested and then system effectiveness assessment, consequence analysis, and risk estimation steps are repeated. The cycle continues until the estimated risk level is estimated to be below the threshold.

12.8   ANALYSIS OF IMPACTS IMPOSED BY RISK REDUCTION UPGRADE PACKAGES

An analysis is completed to evaluate potentially important impacts imposed on the building or facility by the system upgrade packages. Impacts to be evaluated are based on sensitivities and specific site concerns and conditions. Impacts analysis generally estimates impacts on security risk level, costs, operations, schedules, and acceptability by staff or the public. Risk managers need to know and understand important potential impacts imposed by the risk reduction strategies in order to make risk management decisions.

12.9 PRESENTATION TO MANAGEMENT

The final step in the risk assessment process is the preparation of a report of risk assessment results and a presentation package for the risk managers and stakeholders. The analysis team will prepare and present the risk assessment summary. The presentation generally includes the threat description, the security risk estimates for the baseline system, descriptions of any risk reduction packages, and the results of the *impact analysis* for the risk reduction package(s). By comparing this to the baseline risk levels, managers are able to understand what the upgrade package is buying them in risk reduction as well as other potential impacts. The total information package provides invaluable information for risk management decision makers.

12.10 RISK MANAGEMENT DECISIONS

Building owners, stakeholders, and risk managers have the risk assessment information package to help them make difficult security decisions. Several different decision outcomes are possible:

- Accept the security risk level of the baseline system
- Buy more insurance
- Implement one or more risk reduction packages
- Ask analysis team for additional analyses
- Provide contingency measures for security risks that cannot be covered at all times, but can be implemented during periods of heightened threat conditions

Finally, the risk managers decide on the design threat or the threat level to which the security system will be designed. This design threat may be the threat spectrum used in the risk assessment or it may be some subset of that threat spectrum. If the design threat is a subset of the assessment threat spectrum, decisions should be made on a schedule to address the remainder of the threat spectrum as required by threat conditions.

Chapter 13 demonstrates the security risk assessment and management process. This worked example applies the tools and techniques described in this textbook.

Part II

Chapter 13

Demonstration of the Security Risk Assessment and Management Process

13.1 INTRODUCTION

An example building model can provide some of the basic information that is needed for preparing a full security risk assessment process. This building model example will also be applied as the guideline for obtaining pertinent details that are usually important when initiating a security risk assessment and management process of a significant building and its supporting facilities. However, the descriptive categories and the extent of the building details are considered minimal and generic so that the topical areas of an analysis, including a building description and narrative, can form the basis for an initial building survey and characterization when a risk assessment study is required.

The building description details that are provided will focus primarily on the critical assets that are normally evaluated for life-safety and mission-oriented security concerns. Some other building descriptive details that are included in this example might be more informative for subsequent use in future analyses. The building details considered herein as critical assets are those that the management and decision makers are most likely to determine important for now and also for future projects.

This building example will therefore be used to demonstrate the application of the risk assessment and management process contained in this guidebook. A twelve-story building is selected as a hypothetical configuration of a facility that might be considered for a security risk assessment requirement by its owners/stakeholders. The building is located in an urban setting and used for manufacturing expensive jewelry and other similar precious and valuable products. The owner might also not wish to publicly divulge to the community the production process for security reasons. The building also houses the corporate offices and all administrative functions, including a data center, records repository, and inventory control located within the Control Center. The building configuration and its critical assets will be described in a fault tree, and this example will then provide the database for subsequently performing an actual risk assessment and for evaluating some typical risk reduction measures. Any resemblance of this example building to an actual facility in the national inventory is purely coincidental, as this example is only provided as an aid for highlighting the risk assessment and management process using a practical example.

13.2   SECURITY RISK ASSESSMENT AND MANAGEMENT PROCESS

The described systematic analytical process will be used to assess security risks for this example building. Figure 13.1 describes the order and sequence of the basic steps of the process. The process begins with an optional screening analysis for corporations to prioritize their facilities in a business complex. This step will be followed by characterization of the subject building, including identification of the undesired events and the respective critical assets.

Guidance for defining an adversarial threat is included, as well, for applying the definition of the threat spectrum to estimate the threat potential for attack, or likelihood of adversary attack, at this

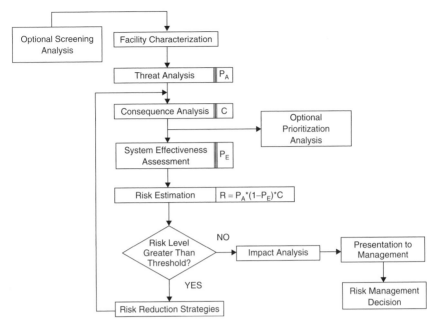

Figure 13.1 Security Risk Assessment and Management Process.

specific building. Relative values of consequence are also estimated for this manufacturing and corporate operation so that any losses from a malevolent attack can be quantified against consequence thresholds established by the owner. An additional and optional step further allows the owner to begin prioritizing the assets at a given facility, if warranted.

Following the application of the threat and consequence assessment analyses, the procedure for estimating the effectiveness of the security system against the adversary attack will be described. Finally, using the three determined parameters of the risk assessment process, the relative risk will be estimated for the example building. In the event that the value of risk is deemed by the owner(s)/stakeholder(s) to be above a predetermined threshold (too *High*), the methodology will address a process for identifying and evaluating risk reduction that would bring security risk to a more acceptable level.

13.3   SCREENING ANALYSIS

Assume that the example building is one of seven buildings owned by a large corporation. The seven buildings are located in various large cities in the country and are multi-use buildings. Several of the buildings are located in high crime areas. The concern is about the impact of a vehicle bomb, theft of valuable assets, or a violent insider incident at one of the buildings. The owners were pondering security concerns and wondered which of the seven buildings required security upgrades and which buildings should be addressed first. A screening analysis was completed to support the decision making.

After lengthy discussions, the conclusions were that the three consequence factors that were most important to the owners were:

- Physical harm to their employees and building occupants
- Loss of the building and contents (economic)
- Loss of operations (economic)

Each of the seven buildings was evaluated for the three factors above in terms of relative consequences of the undesired events. Table 13.1 summarizes the results. Note that the building consequence tally looks like:

- Building A: 2 M, 1 L
- Building B: 2H, 1 M
- Building C: 3H
- Building D: 1H, 2 M
- Building E: 1H, 1M, 1L
- Building F: 3M
- Building G: 2L, 1M

The example building, Building C, has 3 H (*High*) consequence values for loss of lives, loss of building, and loss of operations.

Table 13.1   Consequence Evaluation for Each Building

Building	Undesired Event	Vehicle Bomb	Theft of Assets	Violent Insider
A	Loss of lives	M	L	M
	Loss of building (economic)	M	L	L
	Loss of operations (economic)	L	L	L
	Maximum impact:	M	L	M
B	Loss of lives	H	M	M
	Loss of building (economic)	M	L	L
	Loss of operations (economic)	H	H	L
	Maximum impact:	H	H	M
C	Loss of lives	H	L	M
	Loss of building (economic)	M	M	H
	Loss of operations (economic)	H	H	L
	Maximum impact:	H	H	H
D	Loss of lives	M	L	M
	Loss of building (economic)	H	M	L

(continued overleaf)

Table 13.1 (*continued*)

Building	Undesired Event	Vehicle Bomb	Theft of Assets	Violent Insider
	Loss of operations (economic)	L	L	L
	Maximum impact:	H	M	M
E	Loss of lives	H	L	M
	Loss of building (economic)	L	L	L
	Loss of operations (economic)	L	L	L
	Maximum impact:	H	L	M
F	Loss of lives	M	M	M
	Loss of building (economic)	M	L	L
	Loss of operations (economic)	M	L	L
	Maximum impact:	M	M	M
G	Loss of lives	L	L	M
	Loss of building (economic)	L	L	L
	Loss of operations (economic)	L	L	L
	Maximum impact:	L	L	M

Thus, Building C was placed at the top of the list of buildings to undergo a complete security risk assessment and explains why this security risk assessment is being done.

13.4 FACILITY CHARACTERIZATION

An initial step in a security risk assessment is to characterize the building to be analyzed. Facility characterization requires a thorough understanding of the mission and operating conditions of the building, as well as the security concerns. The security concerns should describe the undesired events – the specific effects and impacts that, ideally, the protection system would be made capable of preventing, and if not preventing, mitigating. An extension of the description of the undesired events is the identification of the organization's critical assets that an adversary would be attempting to harm, destroy, or steal in order to cause undesired events. The valuable material products of this manufacturing facility and the corporate headquarters would be a target for either theft or destruction to cause an economic impact on the owner(s) or nation.

The building characterization procedure must include a complete physical description of not only the configuration and layout of the building but also its construction and operations details. Included in this inventory, at a minimum, are the following: 1) locations of site boundaries and surrounding landscape amenities; 2) layout of all utility services such as electric power, water, and communication; 3) building floor plans and parking facilities that indicate ingress and egress routes for vehicles and pedestrians; 4) property and building access points, including security policy, procedures, and manpower; 5) specific data centers and specialized manufacturing and administrative areas that contain the vital company operations; and 6) all other physical and cyber-protection and security features and their locations.

During the building characterization phase, any known vulnerabilities or weaknesses in protection are noted for future evaluation.

For example, during the interviews with building operators and managers, critical assets are discussed and their functions and protection weaknesses noted. Such discussions might highlight an obvious unprotected exposure of valuable products or materials and possibly the location of uncontrolled access to the building. This example building characterization procedure can be concluded with a general statement of the existing protection system objectives for the facility. Usually, the protection system objectives are based on the list of undesired events or some subset of the undesired events and a summary of the respective critical asset(s) to be protected.

13.5 OPERATIONS

The example building that is used for demonstration of the security risk assessment and management process is the building previously identified in the screening process as requiring a complete risk assessment.

The operational functions of the building include manufacturing space with its vital and expensive equipment, controls, and storage of stock material and finished products. The space is manned 24/7 by operators working 8-hour shifts. Machines are controlled through a programmable system for each product and use the main data center located in the Control Center for reference information and inventory control. The two manufacturing floors (first and second) and the other corporate facilities (third to twelfth floors) meet all safety and health requirements. The manufacturing process is the primary mission and the corporate headquarters is considered the secondary, but critical, functional requirement for the building. Although separated by a distinct secure and operational boundary, the manufacturing areas are integrated into the operation of the corporate and administrative functions by common utilities and by the data center that serves the entire company complex. Except for the two freight elevators within the manufacturing areas that are also used for moving equipment to and from the administrative areas, the other two passenger elevators link all the other non-manufacturing functional areas and originate in the lobby of the first-floor entrance. Two stairwells are located on the north and south side of the building primarily for evacuation and other emergency uses.

Figure 13.2 Example Building.

The corporate and administrative functions are housed in the upper floors beginning with the third floor and continuing to the twelfth. Senior corporate staff members are located in the top two floors (eleventh and twelfth), and access to those floors is by key control from inside the elevators. The administrative functions serve the entire company and also provide support to the national and international offices. Figure 13.2 shows the exterior of the example building.

13.6   GENERAL DESCRIPTION

Off-Site

The building is located in a dual residential and commercial neighborhood that has a sparsely forested setting with easy access to local streets and major highways within a mile. There is a nearby train station for public transportation services to the area and also taxi service. The area is considered a high-crime and high-vandalism district and therefore frequently patrolled by local police. The northern boundary of the facility borders multi-level commercial facilities, many of which have top-story patios that provide clear views of the building. A gate-controlled employee and visitor parking area is provided on the northern boundary. Curbside parking is also available on all sides of the building. The building has a set-back off the curb of approximately 25 feet along the east and west sides and about 50 feet at the south side. The curb-line at the northern edge of the employee parking is about 150 feet from the building because of the parking area and delivery vehicle access.

On the north side of the building where the personnel entrances are located, the building's first floor is at the street level. On the east and west sides of the facility, the respective streets have a gradual downward slope toward the south. The building basement floor is located one floor below the street level at the north side and at street level on the south side. The basement is primarily used for building operations and maintenance functions, storage of high-value stock materials (inside a secure vault room), and for utility supply connections, including water, electric power, gas, and communications.

The north- and east-side boundaries of the building are adjacent to high-rise apartment complexes that have a clear view of the facility through the trees. The west-side boundary of the building faces across the street toward low-rise commercial and office buildings. The southern boundary of the facility is also adjacent to a commercial area. Just a block further south of the building is a commuter train/trolley station and a strip-mall shopping area.

Figure 13.3 shows the example building floor plan and site layout.

Perimeter

Two double-vehicle gates and chain-link fences form the perimeter boundary of the building. One of the vehicle gates is used primarily for employee vehicles; the other is used for delivery and shipment vehicles. Several unsecured manhole covers are located throughout the parking area as access points for the building utilities entering the complex. These are within the fenced area and can be monitored by the security personnel. Access to the building through these manholes is questionable because of the small diameter conduits that are used for the utility lines.

Perimeter Fence

A chain-link fence marks the entire boundary of the building and is mostly along the curb-line. The fence is 6 feet high with double outriggers on top. On the east side of the facility, pine trees hang over a segment of the fence line and could hinder effective surveillance. On the west side, other types of smaller trees are located between the fence-line and curb and some obscuration of the building's first floor is noted. The tautness or strength of

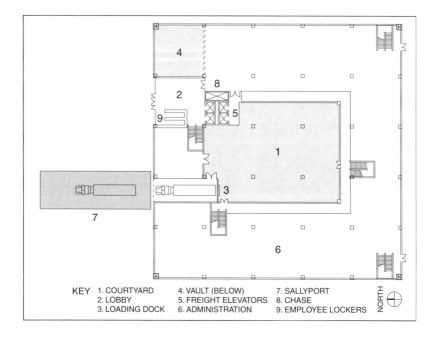

KEY
1. COURTYARD 4. VAULT (BELOW) 7. SALLYPORT
2. LOBBY 5. FREIGHT ELEVATORS 8. CHASE
3. LOADING DOCK 6. ADMINISTRATION 9. EMPLOYEE LOCKERS

NORTH

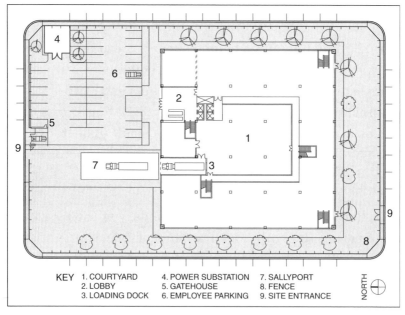

KEY
1. COURTYARD 4. POWER SUBSTATION 7. SALLYPORT
2. LOBBY 5. GATEHOUSE 8. FENCE
3. LOADING DOCK 6. EMPLOYEE PARKING 9. SITE ENTRANCE

NORTH

Figure 13.3 Example Building Floor Plan and Site Layout.

the perimeter fence could be greatly increased by adding additional fence-tensioning wires. Segments of the fence appear to be rusting and deteriorating. The vehicle gates are in a good state of repair and are operated by security officers. Another secured gate is located along the southern boundary for landscape maintenance.

Area Between the Fence and the Building

Closed-circuit television (CCTV) cameras and post-mounted area lights are installed in the area between the fence and the building. Camera images are monitored in the Control Center located on the first floor of the building.

Personnel Vehicle Entrance

The personnel vehicle entrance is lighted and covered by CCTV surveillance. A swinging chain-link gate and a lift bar allow cars to be checked before entrance is granted. Two security police officers staff this gate 24 hours per day, 7 days per week. One officer leaves the structure to check credentials of the driver of the vehicle and to inspect the vehicle by checking in the trunk, under the hood, and under the vehicle using a mirror. The officer inside the masonry shelter operates the swinging chain-link gate and the lift bar. The security guard shelter also has communication and a duress-signaling capability for the officers on duty.

Shipment Vehicle Entrance

The separate vehicle entrance is used primarily for deliveries and shipments. All FedEx and UPS collections and deliveries are also controlled at this vehicle entrance. During shipments or deliveries, one security police officer is dispatched to this gate. The entrance is lighted and is well monitored by CCTV camera surveillance. Delivery vehicles are inspected before the loading dock door is opened by the security officer. The security officer first checks the credentials of the driver to ensure that the driver is listed on the access list. The vehicle is then inspected for contraband. If all requirements are satisfied, the loading door dock is opened, the hydraulic vehicle barriers are lowered, and the vehicle is allowed to proceed to the loading dock ramp inside the building.

Air Route

Although it is not a normal route, it is possible that the adversary could use a hang glider or helicopter to enter the facility boundary. If the hang glider could land on the roof, doors or windows (skylights) on the roof could provide access into the building. Helicopter flights over the building are frequent from a local airport. However, minimal clearance is available on the roof because of rooftop irregularities and obstructions. Landing on the large employee parking area is possible only with prior notification to clear some of the vehicles. Any unexpected landings would be immediately apparent from inside the building. All roof openings and skylights are secured from the inside but are not alarmed.

Building Exterior

The paved driveway along the building's eastern boundary can serve for first responder access and as the temporary emergency evacuation and holding area for employees. Just off the driveway to the east of the building is a tank of nitrogen for use in the manufacturing process. Three hydrogen tanks are also located within the fenced-in area along the same side of the building but about 150 feet further toward the southern side of the building. A major electric power distribution substation is located in the northeast corner of the parking area and is enclosed with a 6-foot concrete-block wall and wrought iron double gate. Several transformers and a small power control and relay building are located inside the enclosure. High-voltage signage is prominently located around the enclosure, and electric safety precautions are well displayed on notices, indicating that electric shock danger is a concern to the company. This enclosure is located in the northeastern quadrant of the vehicle parking area and inside the fenced area to further restrict and protect against public access.

Entrances to Building

The normal entrance is the personnel entrance located just off the sallyport. The courtyard includes various doors that could be used for entrance into the building. Shipments and deliveries employ the vehicle door on the north side of the building. In addition, the emergency exits located on the south side of the building, the various doors from the roof, and some windows could be used to gain entry into the building.

Personnel Entrance

The personnel entrance is located just off the sallyport of the building. The entrance is operational 24 hours per day, 7 days per week. Two security officers staff the entrance. During shift change, two additional officers are present. Employees enter the area and put their personal items in lockers for storage. They then walk through metal detectors. Any items taken into the operations area are screened by the x-ray machine. Upon exiting, hand-carried items and personnel are screened for metal.

Courtyard Entrances

The courtyard at the center of the building contains waste disposal, chemicals, and other hazardous materials. Two 500-kVA electric power backup generators are also located in the courtyard. These generators start automatically and can provide up to 72 hours of backup power without refueling. One of the 10 feet by 10 feet fresh air intake louvers mounted on the building wall is located at ground level in the courtyard; the other is located on the roof of the building. A heating, venting, and cooling (HVAC) chiller is located on the roof for air conditioning for the data center, primarily, and secondarily for the administrative and manufacturing areas.

Entrance into the courtyard from off-site is either through the sallyport or by air approach from the roof. The outer sallyport door is normally open to allow personnel to pass to the personnel entrance. The inner sallyport door is normally closed and locked and is opened remotely by the security officer only to allow the movement of materials into the courtyard or the removal of trash from the courtyard area.

Loading Dock

All deliveries and shipments into and out of the plant are through the single loading dock and are controlled by a tight management security and materials accountability system. The metal rollup door at the loading dock is unalarmed and is opened from the inside. The door is only closed during off hours and very inclement weather.

For shipments of products outside the building, an accountability check is made prior to placing containers on the loading dock. The checks for appropriate shipment, including contents and addresses, are thoroughly reviewed to ensure accuracy and correct inventory of final products. This shipment procedure is also controlled by an inventory process at the data center, and no shipment can be released without authentication by an appropriately authorized individual of the company.

Two emergency exit stairwells empty into the courtyard. The door to the stairwell located in the northwest corner of the courtyard has three glass panels. The door has an emergency panic bar on the inside and is equipped with a balanced magnetic switch (BMS). The metal door to the stairwell located in the northeast corner of the courtyard is unlocked. This door provides access to the boiler room area.

A fresh air intake vent is located in the southwest corner of the courtyard. It is not clear that this route would easily allow access into critical asset areas.

Roof Doors

At the twelfth-floor stairwells from the roof, there are two entrances into the building. These doors are adjacent to the hallways that lead to the corporate offices. The doors are constructed of wood and steel and have glass panels. They are equipped with a BMS, have crash-out bars on the inside, and are key-locked from the outside.

Windows

Various windows on the manufacturing floors (1 and 2) could provide an entry point into the building. However, these windows are equipped with security bars. The windows on the other upper floors are locked from the inside, but not alarmed, and are not equipped with security bars. They open to corporate and administrative offices and are above the manufacturing areas. The windows in the basement are locked and also have security bars. These security bars might be considered a safety hazard and need to be evaluated for that deficiency.

Building Interior

Various routes exist in the interior of the building that lead to critical asset areas. Security features associated with each area will be discussed.

Vault

The high-value hardened vault is located in the building basement at the remote northeast corner, where it is below ground. Exterior access would be from the exterior parking area and courtyard that is secured and monitored. Interior access would be from the interior basement and through the more than 3-inch-thick steel vault door that is mounted on a heavy-duty steel frame and anchoring system. Although access to the vault area is also possible from the courtyard, the exact location is obscured by the floor and partitioned walls. The walls, floor, and ceiling of the vault are all heavily reinforced concrete more than 12 inches thick, and the ceiling and floor are further protected by additional 8-inch-thick concrete slabs that are part of the actual building structure. When the vault is not in use, door switch sensors and penetration sensors are located in the vault door, and motion sensors are located within the vault. CCTV coverage provides an assessment capability. Alarms are annunciated and assessed at the Control Center. Authorized access through the vault door is monitored by security and controlled by the operations managers on duty throughout the 24/7 production periods. Any unauthorized penetration into the vault, using either hand tools or explosives, will be annunciated at the Control Center and will be responded to by an active force.

Control Center

The Control Center is located on the first floor of the building and contains the data center as well as the security alarm annunciator. The control equipment for both the business information technology (IT) system and the production process control system is also located in this center. Security system alarms are annunciated in the Control Center and operators can use a hard-wired telephone to local law enforcement in the event of a security incident. Admittance into the room is controlled by a cipher lock. Only personnel authorized to be in the room are given the combination. Because of the dual-business system and process control operations, at least two operators are on duty at all times.

Utilities

The northeast corner of the first floor level is an important utility node for the entire facility. It houses the steam boilers, the water and gas utility ports, the electric power ports, and the communication node for the facility. The electric power vault and transformer for the building are located just adjacent to this area. This area is noted for potentially being a single point of utility failure and will require careful evaluation as an adversary target.

Security Force

The security force for the building has forty officers. During operational hours, ten officers plus a sergeant and lieutenant are on duty. One officer is located in the Control Center, two at public areas, two at the main vehicle gate, one at the employee entrance, one at the employee exit, one shipment officer, one at the shipping door, an internal patrol, and two exterior patrols in motorized vehicles. Communication is by Motorola Saber digital radios.

Information System

The information system for the example building has a two-level network consisting of the system network and the corporate network. The corporate network controls the business IT system and the production process control system and was designed for protection from outsiders with a strong firewall at the boundary. The system network was designed to support the building activities that need less restricted communication with the Internet. The protection afforded the system network includes a commercial firewall configured with basic protection mechanisms to deny access to important services and servers. An intrusion detection system (IDS) monitors all traffic into and out of both networks.

Undesired Events/Critical Assets

Undesired events result in undesired consequences. Undesired events are site-specific and have adverse impacts on public health and safety, the environment, assets, mission, and publicity. The analysis team met to determine the undesired events for the facility. Natural disasters that could be considered as undesired events were not addressed in this analysis. It is possible that the adversary might use an emergency situation, like an earthquake, to their advantage. Emergency procedures for situations like earthquakes should be addressed.

Four categories of events were identified as security concerns for the facility:

1. Crimes committed against people
2. Damage or destruction of property
3. Disruption of mission
4. Theft of assets

Crimes against people could include hostage situations, murder, or activities causing mass injuries, illness, or casualties. Specific crimes against people were identified and considered a major concern. Sniper fire from the north or east sides of the facility was discussed because of the access provided by the residences and apartments on these sides of the facility. Hostage situations, drive-by shootings, and security force protection incidents were considered. Concerns about a disgruntled employee becoming violent and committing a violent crime against colleagues were also included. Other concerns were mass illness or casualties caused by a hazardous material (chemical or biological agent), contamination of the facility, an explosion caused by the boilers, a bomb brought inside the building, or a vehicle bomb located outside the building.

Destruction of property and loss of high-value assets by explosion, arson, or theft were also considered as undesired events. Aside from the loss-of-mission issues, the monetary values of the building, equipment, materials, or products were considered important.

Building Fault Trees – The identified security concerns and undesired events together with building information were used to create a site-specific fault tree for the example building. The tree can be used to determine critical assets, those that must be protected to prevent an undesired event. The tree can also be used to describe specific scenarios involving critical assets that could produce an undesired event. The modified generic fault tree branches for the example building are pictured in Figures 13.4 through 13.10. The trees are annotated for the specific undesired events listed above.

Critical Assets – Once the undesired events for the facility were established, the next step was to identify the critical assets that

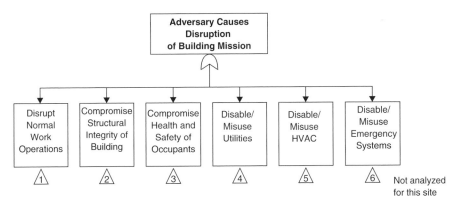

Figure 13.4   Top Level of Generic Building Fault Tree.

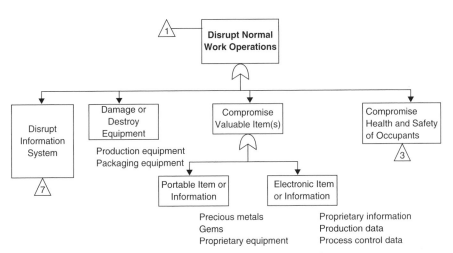

Figure 13.5   Disrupt Normal Work Operations Branch.

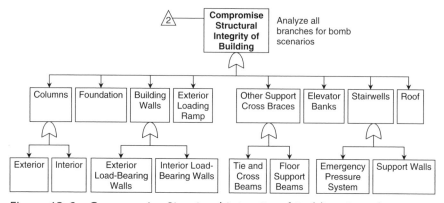

Figure 13.6   Compromise Structural Integrity of Building Branch.

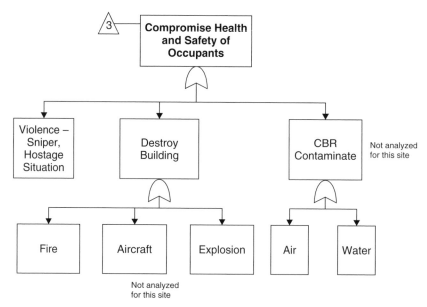

Figure 13.7   Compromise Health and Safety of Occupants Branch.

must be protected in order to prevent the undesired events from occurring. For the four categories of undesired events described above, a set of critical assets was derived. Some assets were common to more than one category of events. Table 13.2 summarizes the undesired events and the associated critical assets

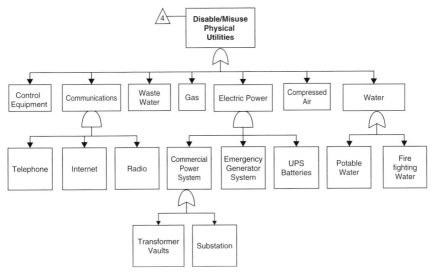

Figure 13.8   Disable/Misuse Physical Utilities Branch.

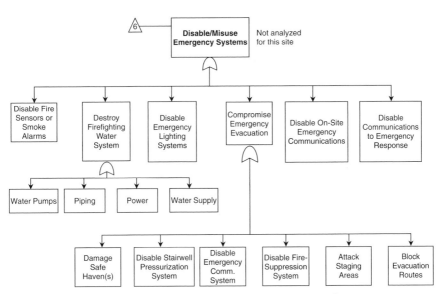

Figure 13.9   Disable/Misuse Emergency Systems Branch.

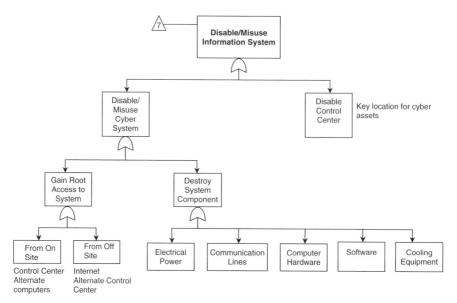

Figure 13.10   Disable/Misuse Information System Branch.

and locations. Because of limited available information, the HVAC system was not fully analyzed in this study. Therefore, the level of mass destruction by chemical/biological agents placed in the air supply was not included. It was noted, however, that the air intake vents on the roof and in the courtyard were not protected. Similarly, hazardous materials were found on-site (nitrogen, hydrogen, other hazardous materials), but the estimation of level of damage or injury that could be caused by releasing or exploding these materials was not part of this study. These materials were judged to be exploitable because of their location on the perimeter of the facility and their lack of protection. The resultant set of critical asset locations were:

• Security and information Control Center
• Exterior grounds/maintenance room/courtyard (utilities/power, wastewater, air intake)
• Parking areas (curbside, parking lots)

Table 13.2   Critical Assets for Building

	Undesired Event	Target	Location
Crime Against People	Sniper	People	Exterior, entrances
	Drive-by shooting	People	Exterior
	Exterior bomb	People/building	Curbsides/ parking areas
	Bomb brought interior	People/assets	Building interior
	Chem/bio attack	People	HVAC
	Hazardous material attack	People	Nitrogen, hydrogen, Haz/mat area
Destruction of Property	Destruction of building/ equipment	Building structure	Curbside/ parking areas, boiler room area, production areas
	Arson	People/building	Building interior
	Tagging	Building/ structures	Exterior walls/ structures
Disruption of Mission	Sabotage/ destroy equipment	Packaging equipment	Production areas

(continued overleaf)

Table 13.2 (*continued*)

	Undesired Event	Target	Location
	Loss of wastewater	Wastewater	Courtyard
	Loss of power	Power	Exterior utility area, interior utility area, sallyport, courtyard
	Cyber-attack	Business IT system	Control Center
		Process control system	On-site computers
	Bomb threat	People	Building
Theft of Assets	Armed robbery	Precious metals, Gems	Vaults
	Proprietary information	Corporate IT system	Control Center
		Process control system	On-site computer terminals

- Production rooms
- Vaults
- Northeast corner of the facility – boilers, water, gas, generators
- Information system

13.7 THREAT

An important parameter of the risk analysis process is the threat potential, particularly the likelihood of adversary attack.

Threat – Before a vulnerability analysis can be completed and before threat potential for attack or likelihood of attack can be estimated, a description of the threat is required. This description includes the types of possible adversaries, tactics, and capabilities (e.g., number in the group, weapons, equipment, and transportation mode).

Threat Potential for Attack (Likelihood of Attack) – After the adversarial threat spectrum has been described, the information is used together with statistics of past events and site-specific perceptions to categorize threats in terms of likelihood that each type of threat would attempt an undesired event. Threat potential is estimated per undesired event and per adversary group. The basis of the parameter estimation is:

- Characteristics of the adversary group relative to the asset to be protected
- Relative attractiveness of the asset to the adversary group

Threat information for the facility was based on the threat data collected and the perceptions of the analytic team. The threat spectrum included outsiders, with the possibility of an insider colluding, and a single insider.

Outsiders

Because of the lack of site-specific data, threat descriptions were examined of historical events with similar concerns. The outsider threat spectrum comprised four basic groups: the terrorist threat, the criminal threat, the extremist threat, and the gang threat. The terrorist threat was defined as two to three individuals armed with conventional weapons (handguns, automatic weapons) and explosives and/or chemical or biological agents. The terrorist threat could also encompass a truck/vehicle bomb attack, backpack of explosives carried in, or an air attack (hang glider, parachute attack). The terrorist threat is capable of cyber-attack(s) on the

IT system for financial purposes or the production process control system to disrupt the mission of the building.

The criminal threat is motivated by financial gain. The threat was described as two to three armed (handguns) individuals intent on stealing materials. They might enter by foot, truck, or air attack and use explosives to enter vaults or protected areas. The criminal threat might consider cyber attack(s) on the IT system or the production process control system.

Extremists or demonstrators are intent on making a political statement. The group would be expected to be relatively few in number. They may or may not resort to violence to achieve their stated goal. Taken to the limit, the extremists resemble the terrorist threat in capabilities.

The final outsider category is the gang member. Neighborhood gangs abound in the area and tend to be organized and violent. A concern is employees becoming gang members. Crimes committed against people, as well as armed robbery attempts were discussed for the gang threat.

Collusion

The collusion threat addresses the possibility that an insider might collude with an outsider threat. This collusion could take the form of the insider being passively involved, providing critical information or knowledge or opening doors, or the insider could be actively participating in the attack.

Insider

The single insider threat was considered for the facility. Personnel undergo a personal background check before employment. Basically, background checks are made with local and regional law enforcement. Follow-on investigations are rarely done. All insider positions that could be held by an insider adversary were considered. The positions were considered in terms of whether or not the position involved authorized access to targets and/or the security system. The insider could be nonviolent (expected to give up if detected) or could be violent (willing to use force to achieve his goal). Motivations for the insider adversary could be disgruntlement with authorities, mental instability, personal financial crisis, anger with coworker, or coercion (family held hostage). Table 13.3 provides a threat spectrum for the facility.

LIKELIHOOD OF ATTACK – Using the above threat statement for the building together with the process described in Chapter 4, the likelihood of attack for the threat spectrum can be estimated. Results of estimating the likelihood of attack for the threat spectrum are summarized in the following tables. Table 13.4 summarizes the terrorist threat, Table 13.5 summarizes the criminal threat, Table 13.6 summarizes the extremist threat, and Table 13.7 summarizes the gang threat.

Table 13.8 summarizes the insider threat, which is analyzed differently from the outsider threats. For the insider threat, a threat severity is estimated. Based on security concerns, the workforce was divided into three categories. These categories were determined by the level of authorized access to critical assets, authority, normal work locations, and access to the security system. It was assumed that all employees, except security officers were screened by the metal detector and x-ray for hand-carried items before entering the production area. Insider categories considered were:

- Production/support
- Business/administration
- Security officers

The production/support category comprised all employees in the Inventory Management, Plant Engineering, Packaging, Quality Assurance, and Management Services Divisions, as well as the

Table 13.3   Threat Definition for Building

Type of Adversary	Number	Equipment	Vehicles	Weapons	Tactics
Terrorist outsider (may include an insider colluding)	2–3	Hand and power tools, body armor, chem/bio	4 × 4, ATV, pickup, aircraft	Handguns, automatics,[2] explosives[3,4]	Cause catastrophic event, hostage, sniper, cyber-attack
Criminal	2–3	Hand tools, body armor	Foot, truck, aircraft	Handgun, explosives	Property theft, cyber-attack
Extremists	5–10	Signs, chains, locks, hand tools	Car, bus	No weapons	Protest, civil disobedience, damage/ destruction
Insider[1]	Single	On-site equipment	Car, pickup, 4 × 4	Handguns, automatics, explosives	Destruction, violence, theft, hacking
Gang	2–6	Hand and power tools, body armor	Car, pickup	Handguns, automatics	Drive-by shooting Robbery Violence

Notes:

[1] Insiders include special interest groups such as employees, contractors, or vendors.

[2] Type of weaponry includes large caliber (.50) automatic weapons with long-range capabilities.

[3] Explosives (hand-carried) would be limited to what one person could carry – pipe bomb (5 lb) or backpack bomb (40–50 lb).

[4] Explosives (vehicle-carried) – compact sedan (500 lb), full-sized sedan (1,000 lb), or van (4000 lb).

Table 13.4 Likelihood of Attack for the Terrorist Threat

Undesired Event:	Capable?	Hist. interest	Hist. attacks	Current interest	Current surveill.	Documented threats	Conseq.	Ideology	Ease of attack	Total	P_A
Adversary Group: Terrorist											
Crimes against People:											
Bomb	Y	5	3	10	6	2	10	10	5	51	H
Haz/mat attack	Y	3	1	2	2	2	10	5	5	30	L
Sniper	Y	3	1	6	6	6	10	10	5	47	H
Chem/bio attack	Y	1	1	6	6	6	10	10	5	45	H
Drive-by shooting	Y	3	1	2	2	2	6	6	5	27	L
Destruction of Property:											
Destruction of bldg./ eqpt.	Y	3	1	2	2	2	6	6	5	27	L
Arson	Y	3	1	2	2	2	10	5	5	30	L
Tagging	Y	3	1	2	2	2	6	6	5	27	L

(continued overleaf)

Table 13.4 (continued)

Undesired Event:	Capable?	Hist. interest	Hist. attacks	Current interest	Current surveill.	Documented threats	Conseq.	Ideology	Ease of attack	Total	P_A
Adversary Group: Terrorist											
Disruption of Mission:											
Cyber attack	Y	1	1	6	6	6	10	10	5	45	H
Sab./dest. eqpt.	Y	3	1	2	2	2	6	6	5	27	L
Loss of wastewater	Y	3	1	2	2	2	6	6	5	27	L
Loss of power	Y	3	1	2	2	2	6	6	5	27	L
Bomb threat	Y	3	1	2	2	2	6	6	5	27	L
Theft of Assets:											
Cyber attack	Y	1	1	10	6	6	10	10	5	49	H
Armed robbery	Y	1	1	2	2	2	5	5	5	23	L

Table 13.5   Likelihood of Attack for the Criminal Threat

	Adversary Group: Criminal									
Undesired Event:	Capable?	Hist. interest	Hist. attacks	Current interest	Current surveill.	Documented threats	Conseq.	Ideology	Ease of attack	Total	P_A
Bomb	Y	3	1	2	2	2	6	6	5	27	L
Haz / mat attack	Y	3	1	2	2	2	6	6	5	27	L
Sniper	Y	3	1	2	2	2	6	6	5	27	L
Chem / bio attack	Y	3	1	2	2	2	6	6	5	27	L
Drive-by shooting	Y	3	1	2	2	2	6	6	5	27	L
Destruction of Property:											
Destruction of bldg. / eqpt.	Y	3	1	2	2	2	6	6	5	27	L
Arson	Y	3	1	2	2	6	5	10	5	34	M
Tagging	Y	3	1	2	2	2	6	6	5	27	L

(continued overleaf)

Table 13.5 (continued)

	Adversary Group: Criminal									
Undesired Event:	Capable?	Hist. interest	Hist. attacks	Current interest	Current surveill.	Documented threats	Conseq.	Ideology	Ease of attack	Total	P_A
Disruption of Mission:											
Cyber-attack	Y	1	1	6	6	6	10	10	5	45	H
Sab./dest. eqpt.	Y	3	1	2	2	2	6	6	5	27	L
Loss of wastewater	Y	3	1	2	2	2	6	6	5	27	L
Loss of power	Y	3	1	2	2	2	6	6	5	27	L
Bomb threat	Y	3	1	2	2	2	6	6	5	27	L
Theft of Assets:											
Cyber-attack	Y	1	1	10	6	6	10	10	5	49	H
Armed robbery	Y	3	1	6	6	6	5	10	5	42	M

Table 13.6 Likelihood of Attack for the Extremist Threat

	Adversary Group: Extremist										
Undesired Event:	Capable?	Hist. interest	Hist. attacks	Current interest	Current surveill.	Documented threats	Conseq.	Ideology	Ease of attack	Total	P_A
Crimes against People:											
Bomb	Y	3	1	2	2	2	6	6	5	27	L
Haz / mat attack	Y	3	1	2	2	2	6	6	5	27	L
Sniper	Y	3	1	2	2	2	6	6	5	27	L
Chem / bio attack	Y	3	1	2	2	2	6	6	5	27	L
Drive-by shooting	Y	3	1	2	2	2	6	6	5	27	L
Destruction of Property:											
Destruction of bldg. / eqpt.	Y	3	1	6	6	6	5	10	5	42	M
Arson	Y	3	1	6	6	6	5	10	5	42	M
Tagging	Y	3	1	2	2	2	6	6	5	27	L

(continued overleaf)

Table 13.6 (continued)

Undesired Event:	Capable?	Hist. interest	Hist. attacks	Current interest	Current surveill.	Documented threats	Conseq.	Ideology	Ease of attack	Total	P_A
Disruption of Mission:											
Cyber attack	Y	3	1	2	2	2	6	6	5	27	L
Sab./dest. eqpt.	Y	3	1	2	2	2	6	6	5	27	L
Loss of wastewater	Y	3	1	2	2	2	6	6	5	27	L
Loss of power	Y	3	1	2	2	2	6	6	5	27	L
Bomb threat	Y	1	1	10	6	6	10	10	5	49	H
Theft of Assets:											
Cyber-attack	Y	3	1	2	2	2	6	6	5	27	L
Armed robbery	Y	3	1	2	2	2	6	6	5	27	L

Table 13.7 Likelihood of Attack for the Gang Threat

Undesired Event:	Capable?	Hist. interest	Hist. attacks	Current interest	Current surveill.	Documented threats	Conseq.	Ideology	Ease of attack	Total attack	P_A
					Adversary Group: Gang						
Crimes against People:											
Bomb	Y	3	1	2	2	2	6	6	5	27	L
Haz/mat attack	Y	3	1	2	2	2	6	6	5	27	L
Sniper	Y	3	1	2	2	2	6	6	5	27	L
Chem/bio attack	Y	3	1	2	2	2	6	6	5	27	L
Drive-by shooting	Y	5	3	10	6	2	10	10	5	51	H
Destruction of Property:											
Destruction of bldg./eqpt.	Y	3	1	2	2	2	6	6	5	27	L
Arson	Y	3	1	6	6	6	5	10	5	42	M
Tagging	Y	5	3	10	6	2	10	10	5	51	H

(continued overleaf)

Table 13.7 (continued)

					Adversary Group: Gang						
Undesired Event:	Capable?	Hist. interest	Hist. attacks	Current interest	Current surveill.	Documented threats	Conseq.	Ideology	Ease of attack	Total	P_A
Disruption of Mission:											
Cyber-attack	Y	3	1	2	2	2	6	6	5	27	L
Sab./dest. eqpt.	Y	3	1	2	2	2	6	6	5	27	L
Loss of wastewater	Y	3	1	2	2	2	6	6	5	27	L
Loss of power	Y	3	1	2	2	2	6	6	5	27	L
Bomb threat	Y	3	1	2	2	2	6	6	5	27	L
Theft of Assets:											
Cyber-attack	Y	3	1	2	2	2	6	6	5	27	L
Armed robbery	Y	3	1	2	2	6	5	10	5	34	M

Table 13.8 Insider Threat Severity (TS)

Job Category	Access to Physical Assets	Access to Cyber-Assets	Authority	Access to Security System	Opportunity for Collusion	Threat Severity
Production/support						
Supervisor	High	High	High	Low	High	H
Technician	High	High	High	Low	High	H
Business/administration						
System administrator	Low	High	High	Low	High	H
Security officers	Low	Low	High	High	High	H
Security maintenance	Low	Low	Medium	High	High	M

Office of the Superintendent. This category of insiders was assumed to have access to physical critical assets during some part of their job assignments.

The business/administration category comprised the Budget and Accounting, Equal Employment Opportunity, Human Resources, and Information Technology Divisions. The job assignments of this category would not normally provide direct access to physical critical assets but would provide access to the information systems (corporate IT and production process control).

The security force category included all security officers authorized to staff a security post, provide random patrol of the facility, or access the physical protection system components. These assignments may or may not include access to production critical assets.

Four undesired events were analyzed for the insider adversary threat:

1. Violence (in the workplace)
2. Destruction of mission-critical equipment (sabotage, fire)
3. Planting a bomb or device inside the facility
4. Cyber-attack on the business IT system or production process control system.

13.8   CONSEQUENCES

The second parameter of security risk is consequence. Consequence analysis can be completed after the undesired events and associated critical assets have been identified as a part of facility characterization. The next analysis step is to estimate consequences associated with the loss of specific critical asset(s) for each undesired event. Table 13.9 provides the example Consequence Definition Table that was used to estimate relative consequence levels.

The analysis team considered the undesired events and categorized them into consequence categories based on their perceptions of consequences of loss. Table 13.10 summarizes the team's estimation of consequence level for undesired events.

Table 13.9 Consequence Definitions

Consequence Category	Consequence Level
Total collapse of building structure Economic loss greater than $1 million Operations downtime one or more years Security incident resulting in grave damages to corporate reputation	High
Damage to building structure but no collapse Economic loss greater than $500 thousand but less than $1 million Operations downtime – months Security incident resulting in moderate impact on corporate reputation	Medium
Little or no damage to building structure Economic loss less than $500 thousand Operations downtime – hours to days Security incident resulting in minor impact on corporate reputation	Low

The results of the consequence estimation exercise are listed below.

High:

1. Bomb – exterior, interior
2. Chemical/biological attack
3. Arson (fire that results in loss of life)
4. Sniper
5. Violent employee
6. Armed robbery that results in loss of life
7. Drive-by shooting
8. Loss of wastewater

(This "results list" is continued on page 238.)

Table 13.10   Estimation of Consequence Level for Undesired Events

Consequences of Undesired Events				
	Measure of Consequence		Consequence Severity	
Undesired Event	*Type*	*Value*	*By Type* *H / M / L*	*By Event* *H / M / L*
Bomb – exterior or interior	Building collapse	*High*		
	Economic loss	*High*		
	Downtime	*High*		
	Security incident impact- reputation	*Medium*		
			Enter highest consequence	H
Hazardous material incident	Building collapse	*Low*		
	Economic loss	*High*		
	Downtime	*Medium*		
	Security incident impact- reputation	*High*		
			Enter highest consequence	H

Table 13.10   (*continued*)

Consequences of Undesired Events				
	Measure of Consequence		*Consequence Severity*	
Undesired Event	*Type*	*Value*	*By Type* *H / M / L*	*By Event* *H / M / L*
Chem/bio incident	Building collapse	*Low*		
	Economic loss	*High*		
	Downtime	*Medium*		
	Security incident impact-reputation	*Medium*		
			Enter highest consequence	H
Arson (large fire)	Building collapse	*Medium*		
	Economic loss	*High*		
	Downtime	*High*		
	Security incident impact-reputation	*Medium*		
			Enter highest consequence	H

(*continued overleaf*)

Table 13.10 *(continued)*

Consequences of Undesired Events				
	Measure of Consequence		*Consequence Severity*	
Undesired Event	*Type*	*Value*	*By Type* H / M / L	*By Event* H / M / L
Sniper	Building collapse	*Low*		
	Economic loss	*Medium*		
	Downtime	*Low*		
	Security incident impact-reputation	*High*		
			Enter highest consequence	H
Violent employee	Building collapse	*Medium*		
	Economic loss	*Low*		
	Downtime	*Low*		
	Security incident impact-reputation	*High*		
			Enter highest consequence	H

Table 13.10 (continued)

Consequences of Undesired Events				
	Measure of Consequence		Consequence Severity	
Undesired Event	Type	Value	By Type H / M / L	By Event H / M / L
Armed robbery with casualties	Building collapse	Low		
	Economic loss	Medium		
	Downtime	Low		
	Security incident impact-reputation	High		
		Enter highest consequence	H	
Drive-by shooting	Building collapse	Low		
	Economic loss	Medium		
	Downtime	Low		
	Security incident impact-reputation	High		
		Enter highest consequence	H	

(continued overleaf)

Table 13.10　(*continued*)

Consequences of Undesired Events				
	Measure of Consequence		Consequence Severity	
Undesired Event	*Type*	*Value*	*By Type* *H / M / L*	*By Event* *H / M / L*
Loss of wastewater	Building collapse	*Low*		
	Economic loss	*Medium*		
	Downtime	*High*		
	Security incident impact-reputation	*Low*		
			Enter highest consequence	H
Destruction of building	Building collapse	*High*		
	Economic loss	*High*		
	Downtime	*High*		
	Security incident impact-reputation	*High*		
			Enter highest consequence	H

Table 13.10 *(continued)*

Undesired Event	Measure of Consequence Type	Value	Consequence Severity By Type H/M/L	By Event H/M/L
	Consequences of Undesired Events			
Loss of power	Building collapse	*Low*		
	Economic loss	*Medium*		
	Downtime	*Low*		
	Security incident impact-reputation	*Low*		
			Enter highest consequence	M
Cyber-attack – business IT system	Building collapse	*Low*		
	Economic loss	*Medium*		
	Downtime	*Low*		
	Security incident impact-reputation	*Medium*		
			Enter highest consequence	M

(continued overleaf)

Table 13.10 (*continued*)

Consequences of Undesired Events				
	Measure of Consequence		Consequence Severity	
Undesired Event	*Type*	*Value*	*By Type* H / M / L	*By Event* H / M / L
Sabotage of equipment	Building collapse	*Low*		
	Economic loss	*Medium*		
	Downtime	*Medium*		
	Security incident impact-reputation	*Medium*		
			Enter highest consequence	M
Cyber-attack – process control system	Building collapse	*Low*		
	Economic loss	*Medium*		
	Downtime	*Medium*		
	Security incident impact-reputation	*Medium*		
			Enter highest consequence	M

Table 13.10   *(continued)*

Consequences of Undesired Events				
	Measure of Consequence		Consequence Severity	
Undesired Event	*Type*	*Value*	*By Type* $H / M / L$	*By Event* $H / M / L$
Bomb threat	Building collapse	*Low*		
	Economic loss	*Low*		
	Downtime	*Low*		
	Security incident impact- reputation	*Low*		
			Enter highest consequence	L
Tagging	Building collapse	*Low*		
	Economic loss	*Low*		
	Downtime	*Low*		
	Security incident impact- reputation	*Low*		
			Enter highest consequence	L

Medium:

1. Sabotage of equipment (partial loss of operations)
2. Loss of power
3. Cyber-attack on business IT system
4. Cyber-attack on production process control system

Low:

1. Tagging
2. Bomb threat

13.9   PRIORITIZATION ANALYSIS

Tables 13.11 through 13.14 consider the likelihood of attack in conjunction with the consequence values associated with the undesired events for the terrorist, the criminal, the extremist, and the gang threat, respectively.

Undesired Events for Analysis – Based on the tables above, the undesired events for each threat were prioritized. Those undesired events that were associated with *Medium* or higher likelihood of attack and *Medium* or higher consequence values for a given threat were selected for further analysis.

Terrorist Threat

- Sniper
- Bomb
- Chemical/biological attack
- Arson
- Cyber-attack – mission, theft

Table 13.11 Likelihood of Attack vs. Consequence for the Terrorist Threat

	Low Likelihood	Medium Likelihood	High Likelihood
High Consequence	Drive-by shooting Armed robbery Loss of wastewater Hazardous material attack Arson		Bomb Chemical/biological attack Sniper
Medium Consequence	Sabotage equipment Loss of power Destruction of building/equipment		Cyber-attack – mission Cyber-attack – theft
Low Consequence	Tagging Bomb threat		

Table 13.12   Likelihood of Attack vs. Consequence for the Criminal Threat

	Low Likelihood	Medium Likelihood	High Likelihood
High Consequence	Bomb Hazardous material attack Sniper Chem/bio attack Drive-by shooting Destruction of building/equipment Loss of wastewater	Arson Armed robbery	
Medium Consequence	Sabotage equipment Loss of power		Cyber-attack – mission Cyber-attack – theft
Low Consequence	Tagging Bomb threat		

Table 13.13   Likelihood of Attack vs. Consequence for the Extremist Threat

	Low Likelihood	Medium Likelihood	High Likelihood
High Consequence	Bomb Hazardous material attack Sniper Chem/bio attack Drive-by shooting Destruction of building/equipment Arson Armed robbery Loss of wastewater		
Medium Consequence	Sabotage equipment Loss of power Cyber-attack – mission Cyber-attack – theft		
Low Consequence		Tagging	Bomb threat

Table 13.14 Likelihood of Attack vs. Consequence for the Gang Threat

	Low Likelihood	Medium Likelihood	High Likelihood
High Consequence	Bomb Hazardous material attack Sniper Chem/bio attack Destruction of building/equipment Loss of wastewater	Arson Armed robbery	Drive-by shooting
Medium Consequence	Sabotage equipment Loss of power Cyber-attack – mission Cyber-attack – theft		
Low Consequence	Bomb threat		Tagging

Criminal Threat

- Arson
- Armed robbery
- Cyber-attack – mission, theft

Extremist Threat

- None

Gang

- Drive-by shooting
- Arson
- Armed robbery

Table 13.15 summarizes the results of the prioritization analysis, especially the undesired events that were judged to have *Medium* or higher consequences and *Medium* or higher likelihood of attack for one or more adversary groups. Next, protection system effectiveness will be assessed for preventing these undesired events.

13.10   PROTECTION SYSTEM EFFECTIVENESS

The third parameter in assessing security risk, system ineffectiveness $(1 - P_E)$, can be derived from a security system effectiveness assessment. Security system ineffectiveness (adversary success) and security system effectiveness (P_E) are complementary functions. If security system effectiveness is *High*, then security system ineffectiveness (adversary success) is judged to be *Low*. The risk

Table 13.15 Prioritized Undesired Events for Analysis

	Sniper	Drive-by Shooting	Bomb	Chem/bio Attack	Haz/mat Attack	Arson	Destroy Bldg./Eqpt.	Cyber-Attack – Mission	Cyber-Attack – Theft	Armed Robbery
Terrorist	X		X	X				X	X	
Criminal						X		X	X	X
Extremist										
Gang		X				X				X

assessment process will evaluate security system effectiveness in order to estimate system ineffectiveness (adversary success).

For most applications, a security system is made up of physical protection features and cyber-protection features. Some undesired events can be accomplished by a physical attack on the facility while others can be accomplished by a cyber-attack on the system. A total security system should address both physical and cyber-attacks, as appropriate. A complete system effectiveness assessment will include a physical protection analysis and cyber-protection analysis.

13.10.1   Physical Protection System Effectiveness

An effective physical protection system (PPS) must be able to detect the adversary early enough and delay the adversary long enough for the security response force to arrive and neutralize the adversary before the mission is accomplished. In particular, an effective PPS provides effective detection, delay, and response. These physical system functions (detection, delay, and response) must be integrated to ensure that the adversarial threat is neutralized before the mission is accomplished.

Adversary Sequence Diagrams (ASDs) were used to estimate protection system effectiveness for the threats and undesired events derived previously. For some of the undesired events, such as the exterior bomb, drive-by shooting, and sniper cases, the adversary does not really interact with the physical protection system of the facility and an ASD was not developed for these events. The ASD is a graphical representation of physical protection elements along paths that adversaries can follow to accomplish their objective. For a specific physical protection system and threat, the most vulnerable path can be determined. This path with the least physical protection system effectiveness establishes the effectiveness of the total physical protection system. An ASD is developed for a single critical asset associated with a specific undesired event.

The analyses were run with a range of security response force times from 0 seconds to 180 seconds. Operating Conditions 1 and 2 were considered for the analysis. Condition 1 was the production hours when building operations were under way. Condition 2 described operating conditions during the nonproduction hours of the day or week. Emergency conditions (fire alarm) and shipments were evaluated as a part of Condition 1.

Offsite Attacks – Exterior Bomb (Terrorist Threat), Sniper, Drive-By Shooting

Exterior Bomb (terrorist threat):

This undesired event could be accomplished by a vehicle bomb detonated outside the building. The vehicle bomb is assumed to be a vehicle carrying approximately 500 pounds of TNT-equivalent explosives. The targets are vehicle locations adjacent to the building. Results address the estimated success of the adversary to drive a vehicle carrying explosives near the building or park the vehicle adjacent to the building and detonate the explosives before the incident could be interrupted. System effectiveness for this scenario is judged to be low because public vehicles are allowed to park adjacent to the building. A second type of result describes the damage or consequence that is incurred if the adversary is successful in detonating the explosives. The latter results are presented in the blast effects Section 13.10.3.

Sniper (Terrorist Threat) and Drive-By Shooting (Gang Threat):

For these undesired events, the human target could be any employee – a production worker, a security officer, or a building occupant who happens to be outside the building. The north and east sides of the building are exposed to neighboring buildings that could harbor a sniper. Traffic on adjacent streets to the building is moderate. The security system does not affect such scenarios. Because employees are outside the building at regular times, the system effectiveness values for these scenarios are judged to be at the *Low* level.

Armed Robbery (Criminal Threat, Gang Threat)
Several possible targets of armed robbery exist at the building. Vaults are located on several floors including the basement. For this analysis, Vault L was used for the target because it is known to occasionally have a shipment quantity of material and because of its proximity to the shipping/receiving door. Insider collusion could provide adversaries with information about shipment days and times. Further, during shipments, the shipping/receiving door and the vault door could be open at the same time. It is assumed that a security officer is posted at the shipping/receiving door during shipments. The ASD for the vault is shown in Figure 13.11.

Adversary Scenario: Adversaries climb over or bridge fence, traverse to shipping/receiving door that is open during a shipment, neutralize security officer, enter building and vault, pack up assets, and exit the same way that they entered.

A major system vulnerability exists for the detection system in alarm display and assessment operations. This weakness affects numerous adversary scenarios. The main sources of detection on the perimeter are the camera surveillance and the security force – no exterior sensors are present. Poor lighting and camera output result in poor images on the monitors for the security officer in the Control Center. In addition, it is not possible for the security officers in the Control Center to effectively watch 20 or more monitors to detect an anomaly. The video replay for the alarm system was not timely enough to be effective. The security force cannot provide reliable detection because they are normally in an unprotected position, and they do not all have a duress signaling capability.

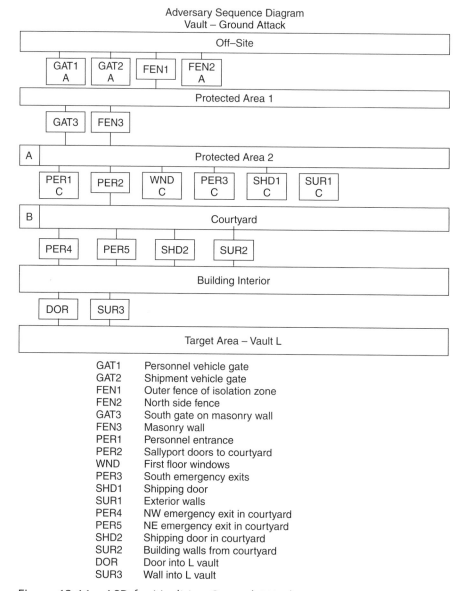

Figure 13.11   ASD for Vault L – Ground Attack.

Discussion: For Condition 1, the system effectiveness was judged to be *Low* because of deficiencies in the detection and delay functions. Because of the deficiencies in the detection system discussed earlier and the lack of delay time with the doors open, system effectiveness is judged to be very *Low*.

Discussion: For Condition 2, it was assumed that the shipping/receiving door and vault doors were both closed and locked. The lack of an effective intrusion detection system caused the system effectiveness to be *Low*.

Courtyard Attacks – Plant Interior Bomb (Terrorist Threat), Arson (Terrorist, Criminal, Gang Threats), Chem/Bio Agents in Air Intake (Terrorist Threat)

Several of the undesired events for the building can be accomplished by accessing the courtyard. These undesired events could result in loss of mission, loss of valuable property, and loss of human lives.

A bomb planted in the building interior (in the courtyard) could cause major destruction. Blast effects are described in Section 13.10.3. The northeast corner of the facility is critical because of the location of the boilers and the utilities – gas, electrical, and water. Both safety (explosion) and loss of mission issues are a great concern.

One of the fresh-air intakes for the building is located in the courtyard; the other is located on the roof of the building. The undesired event of causing casualties or illness by a chemical/biological agent introduced into the air system was not completely addressed in this analysis. The ability of the adversaries to reach the air intake vent before being interrupted by the security force has been addressed. It is feasible that an aerosolized canister of a chemical agent could be dispersed into the air intake vent. If this undesired event becomes a concern, a consequence analysis should be completed. This analysis would address a spectrum of chemical agents and concentrations, various methods and locations of

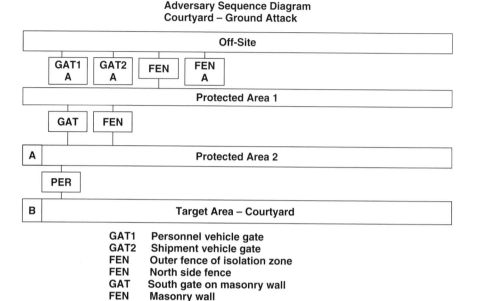

Figure 13.12   ASD for Courtyard – Ground Attack.

introducing agents into the HVAC system, the human physical responses to the agents, and a thorough characterization of the HVAC system.

The ASD for these adversary scenarios is shown in Figure 13.12.

Adversary Scenario: Adversaries climb over or bridge the fence, enter the courtyard via the sallyport doors, complete their task. It is assumed that the adversaries are successful if they can complete the task at the target before local law enforcement arrives.

Discussion: The physical protection system effectiveness was judged to be *Low* because of deficiencies in the detection and delay functions. General deficiencies in the detection function were discussed earlier. Only a single locked door controls access into the courtyard. This does not provide sufficient delay to the adversary for the security response force to arrive.

Information System – Physical Attack

Loss or compromise of the information system can occur via a cyber-attack or by a physical attack, especially to the Control Center. The Control Center is located on the first floor of the building and contains the control equipment for both the business IT system and the production process control system. Security system alarms are also annunciated in the Control Center and operators can use a hard-wired telephone to local law enforcement in the event of a security incident. Admittance into the room is controlled by a cipher lock. Only personnel authorized to be in the room are given the combination. Because of the dual-business system and process control operations, at least two operators are on duty at all times. The ASD for the Control Center is provided in Figure 13.13.

Adversary Scenario: Adversaries climb over or bridge fence, traverse to building door that is open by use, neutralize security officer, enter building, and proceed to Control Center. They enter the Control Center door by force and damage/destroy Control Center equipment. The cyber (electronic) scenario will be addressed later.

Discussion: For Condition 1, the system effectiveness was judged to be *Low* because of deficiencies in the detection and delay functions. Because of the deficiencies in the detection system discussed earlier and the lack of delay time with the doors open, system effectiveness is judged to be very *Low*.

Discussion: For Condition 2, it was assumed that the building doors were closed and locked. The lack of an effective intrusion detection system caused the system effectiveness to be *Low*.

Insider

An insider employee becoming violent or deciding to damage or destroy is extremely difficult to predict. Background investigations may or may not be valuable in predicting such an outcome. If an employee decides to become violent, he or she could injure or kill

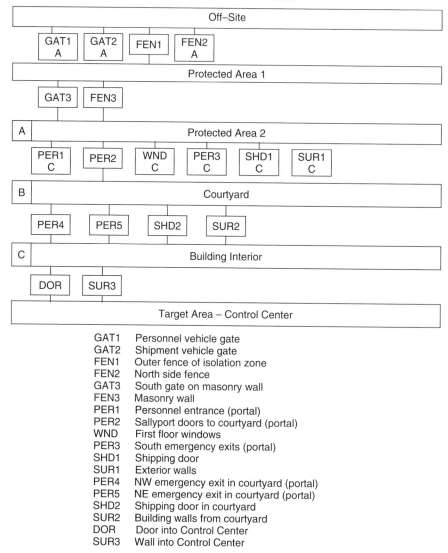

Figure 13.13   ASD for Control Center.

coworkers, damage or destroy mission-critical equipment, or cause mass destruction and death with a bomb planted inside the facility.

Adversary Scenario: All three of the undesired events described above for the insider adversary threat require common tactics. Three possible avenues exist for the insider to accomplish the undesired events:

- Contraband (weapon, flammables, explosives) brought in through the personnel entrance
- Contraband (weapon, flammables, explosives) brought in through other routes
- Contraband (tools, hazardous materials) exists or could be covertly constructed into weapon or bomb on-site

Scenario A. The insider adversary would attempt to bring in the contraband (tools, weapon, flammables, explosives) that he or she planned to use to accomplish the goal. It was assumed that the system effectiveness against this scenario was based on the probability of detection of the contraband. If the adversary were detected, he would give up or be neutralized by the security officers at the personnel entrance. The probability of detection was based on the detection of metal guns or tools. The probability of detection of a gun was estimated to be *High* for the metal detector and package x-ray machine for employees in the production/support and business/administration categories and *Low* for members of the security force who can bypass the screening process.

Insider Scenario A	Metal Det.	X-Ray
Production/support	High	High
Business/administration	High	High
Security officer	Low	Low

Scenario B. The insider adversary would attempt to bring in the contraband (tools, weapon, flammables, explosives) that he planned to use to accomplish his goal through other than normal routes. One way might be to arrange to have the contraband hidden in normal supplies that enter the site at either the courtyard dock or the shipping door. The assumptions made are that all incoming supplies are thoroughly inspected for contraband. Further, it was assumed that the inner sallyport door was strictly monitored when open to ensure that someone could not enter, thereby bypassing the screening point, or use this route to retrieve contraband stashed outside the screening point and bring it into the courtyard. The system effectiveness estimate against this scenario was based on the probability of detection of the contraband. If the adversary were detected, he would give up or be neutralized.

Insider Scenario B	P_E
Production/support	High
Business/administration	High
Security officer	Low

Scenario C. The insider adversary would acquire or construct a weapon, flammable, or bomb inside the production area to accomplish his goal. It was assumed that the system effectiveness against this scenario was based on the likelihood of detection of adversary action by personnel. Two categories of insiders would not be expected to routinely be in the production areas.

Insider Scenario C	P_E
Production/support	Low
Business/administration	NA
Security officer	NA

Insider – Theft of Critical Assets (Physical)

For this undesired event, two adversary scenarios were considered.

1. Adversary obtains critical asset, plants them in waste or recycling, and retrieves them later. All materials taken to the courtyard as waste or recycling should be screened.
2. Adversary activates fire alarm, takes critical asset out during evacuation, and stows them for later retrieval. Even though procedures call for security officer to monitor employees during evacuation to the east side of the building, the insider adversary could arrive at the holding area before the security force and either stow the stolen item(s) for later retrieval or throw it (them) over the fence.

Because there are no definitive protection features to detect, delay, or respond to the production/support insider, the physical protection system effectiveness level was judged to be *Low* for both scenarios. The other two categories of insiders, business/administration and security officer are not expected to have access to the production areas or vaults where finished critical assets would be located.

Summary of Physical Protection System Effectiveness

Table 13.16 summarizes the effectiveness levels of the physical protection system for the adversaries/scenarios.

So far the focus has been on physical protection system effectiveness assessment. A valuable product of assessing system effectiveness is the identification of specific vulnerabilities of the protection system. If the security system effectiveness is judged to be *Low*, specific weaknesses and the associated deficient protection elements causing the *Low* level are site-specific system vulnerabilities. Knowledge of site-specific vulnerabilities is valuable for

Table 13.16 Summary of Physical Protection System Effectiveness for Baseline System

Adversary	Terrorist	Criminal	Gang	Insider		
				Production / Support	Business / Administration	Security Officer
Undesired event:						
Sniper	Low					
Bomb – exterior	Low					
Bomb – interior	Low					
Chem/bio attack	Low					
Arson		Low	Low			
Physical attack – cyber eqpt.	Low	Low				
Armed robbery		Low	Low			
Drive-by shooting			Low			
Insider – violence				Low	High	Low
Insider – sab. eqpt				Low	High	Low
Insider – bomb				Low	High	Low
Insider – theft of item				Low	N/A	N/A

planning system upgrades to reduce risk and for contingency planning to know where to place reinforcement protection during times of elevated threat conditions.

The protection system for the building demonstrated *Low* protection effectiveness for the terrorist, criminal, gang, and insider threats. An effective security system demonstrates a *High* performance level for the detection, delay, and response functions. The protection system for the building demonstrated weaknesses in all three functions. A major system vulnerability exists for the detection system in alarm display and assessment operations. This weakness dominated numerous adversary scenarios. The main sources of detection on the perimeter are the camera surveillance and the security force. (No exterior sensors are present.) Poor lighting and camera output result in poor images on the monitors for the security officer in the Control Center. It is unrealistic to expect the security officer in the Control Center to effectively watch 20 or more monitors to detect an abnormality. The video replay for the alarm system was not timely enough to be effective. Detection of the insider removing valuable items from the building was suspect for some scenarios. System delay was marginal, at best, for some scenarios because critical doors were open and the security force was not in hardened (protected) positions. Response effectiveness was *Low* because response times were too long in most cases.

The northeast corner of the building is an area of great concern. Most of the utilities – water, gas, electrical power, including backup – for the facility could be cut off from this area. An explosion in the boiler room might destroy much of the building, including critical production areas. Hazardous material tanks are located nearby. It is important to keep adversaries out of this area.

Multiple vulnerable paths were found for each undesired event, and many of the vulnerable paths had common elements. The common elements are described below.

Detection

- No detection due to lack of sensors, means of assessment, overload in Control Center
- No CCTV coverage for some door alarms
- No sensors for roof landings or traversals
- Security force used for detection, but some lack body armor, protected positions, duress capability
- Trees on perimeter challenge assessment capability
- Lack of screening of insiders during evacuations
- Lack of detection of insider removal of critical assets from building for specific scenarios

Delay

- Doors open (shipping door and vault during shipments, production rooms)
- Critical utilities and storage of hazardous materials are street-accessible and unprotected (power generators, hydrogen, nitrogen)
- Buildings near fence allow for bridging

Response

- Lack duress capability for all officers
- Lack backup communication
- Lack body armor and protected positions (especially shipping dock)
- Quicker response time required for some scenarios

Cyber-Protection System Effectiveness

Much like an effective PPS that demonstrates *High* performance for the three functions of detection, delay, and response and the integration of these functions, an effective cyber-protection system demonstrates *High* performance for three basic cyber-security functions and their integration. These functions are used to ensure the properties of confidentiality, integrity, and availability of data. *Confidentiality* requires that information not be made available to unauthorized individuals, entities, or processes. *Integrity* requires that information not be altered or destroyed in an unauthorized manner. *Availability* requires that information be accessible and usable on demand by an authorized entity. The three cyber-protection functions are:

- Authentication
- Authorization
- Audit

Authentication, authorization, and audit must be performed at a high level and must be integrated. The authentication and authorization strategies both provide data to the audit capability where it is analyzed for evidence of malicious activity.

Cyber-Attack – Loss of Mission / Theft of Assets (Terrorist, Criminal, Insider Threats)

Various scenarios exist for cyber-attacks on the business IT system to cause loss of mission or loss of proprietary information or attacks on the process control system to cause loss of mission. Figure 13.14 pictures the architecture for the information system for the building. The information system has a two-level network, consisting of the system network and the corporate network. The corporate network controls the business IT system and the production process control system.

Because the business IT system and the production process control system are configured and protected identically, the

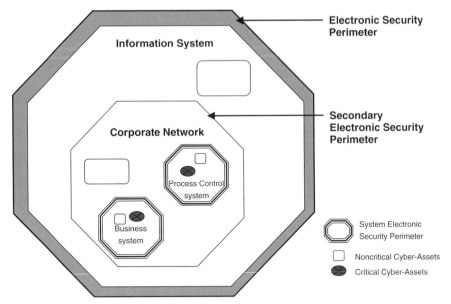

Figure 13.14 Building Information System Architecture.

cyber-path-diagram featured in Figure 13.15 is appropriate for critical cyber-assets in either system.

Scenarios:

Outsider threat – The outsider threat uses the Internet to attempt to gain access to the electronic security perimeter system of the information system, the secondary electronic security perimeter of the corporate network, and either the business IT system or the production process control system to reach the critical cyber-asset(s) to cause loss of mission or loss of proprietary information.

Insider threat – The insider threat uses his or her individual authorized access level to access the corporate network and either the business IT system or the production process control system

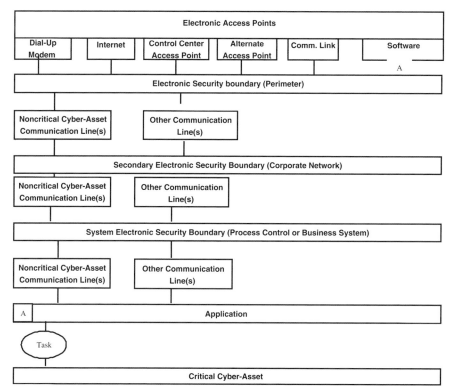

Figure 13.15   Cyber-Path-Diagram for Critical Cyber-Assets in Business IT System or Production Process Control System.

to cause disruption in mission or loss of proprietary information. Three categories of insider positions were analyzed:

- Production/support or business/administration position
- Computer system administrator position for either the business IT system or the production process control system
- Security officer position

Table 13.17 summarizes the building's protection features for the cyber-functions: authentication, authorization, audit, and integration.

Table 13.17 Cyber-system Protection Features

Authentication	Authorization	Audit	Integration
Passwords – user-defined, no system requirements	All employees and contractors have access to information system.	Intrusion detection system at perimeter electronic boundary	Firewall at electronic security perimeter (information system boundary)
	All business/ administration staff are on access list for business IT system.	Scanners Virus protection Access control monitoring	Firewall at secondary electronic security boundary (corporate network)
	All production/ support staff are on access list for production Process control system.	Random traffic data review	No encryption
	Computer system administrators have access to all systems and all assets.		

For the outsider threat, the cyber-protection-system effectiveness for causing loss of mission by attacking the business IT system or the production process control system or causing loss of sensitive information by attacking the business IT system was judged to be at the *Low* level of effectiveness. The minimum value of the assessments for the authentication, authorization, audit, and integration functions defined the overall cyber-protection effectiveness level. Authentication was judged to be *Low* effectiveness because of the weak passwords; authorization was judged to be of *Low* effectiveness because of the coarse groupings in the access lists; audit was judged to be of *High* effectiveness; integration was judged to be of *Medium* effectiveness because the systems have firewalls but no encryption features.

For the insider threat categories, the computer system administrator position posed the greatest threat to the cyber-system because of the authorized access to systems. Cyber-protection-system effectiveness was judged to be *Low* for the system administrator category; system effectiveness was judged to be *Low* for the production/support or business/administration positions for the same reasons listed above for the outsider threat; the security officer position is not authorized to access the information system.

Summary of Cyber-Protection-System Effectiveness

Table 13.18 summarizes the cyber protection system effectiveness assessments for the threat spectrum.

Table 13.18   Cyber-Protection-System Effectiveness Assessments

	Loss of Mission	*Loss of Sensitive Information*
Terrorist	*Low*	*Low*
Criminal	*Low*	*Low*
Business/administration or production/support employee	*Low*	*Low*
Computer system administrator	*Low*	*Low*
Security officer	NA	NA

The site-specific vulnerabilities for the cyber protection system include:

Authentication

- Weak passwords (user sets own)
- Only single-factor authentication into critical asset areas

Authorization

- Access list groupings too coarse (too many categories given access to systems)
- No protection or restrictions for computer system adminis-trator access

Audit

- Data reviews could be done more often

Integration

- No encryption features
- No firewalls to separate business and process control sys-tems

13.10.2   Analysis of Blast Effects

Existing Building Configuration

The building is a steel-frame structure with thick concrete bearing-wall panels securely mounted to the structure to form the exterior envelope. The four corners of the building are supported by the main steel and concrete-covered columns that are the key structural elements of the building. All of the steel floor beams are connected to these columns, but some of the floor slabs are poured in place or precast and are freely resting on these beams. These floor slabs do not support any upward loads that would be caused by an explosive airblast. The columns are tied together through the floor beam array and by additional diagonal steel elements on selected floors to minimize any potential twisting caused by wind and earthquake loads. These four composite building columns are also embedded in deeply buried concrete pier foundations that are supported by competent bedrock. However, any severe damage to any one of the four columns, especially at or just above the basement or the first floor level, will weaken the structure and result in progressive building collapse and major disruption to the numerous interior and exterior utility systems, including electric power, water-wastewater, gases (nitrogen and hydrogen), HVAC, and communications.

The steel roof beams help support the roof's open-web steel joists. The roof consists of a metal deck, lightweight-insulating concrete, and a built-up roof system. The mechanical and electrical equipment penthouse area, located on the surface of the roof, is a constructed medium-sized steel-stud-framed building with steel roof joists. Radio and microwave antenna and communication systems are also mounted on the roof deck. Access doors and utility penetrations on the roof are provided for enhanced lighting on the twelfth floor, utilities, and for personnel access during operations and maintenance periods.

The exterior wall consists of a variety of cementitious materials, including reinforced concrete panels, granite sills, insulation, and interior drywall or plaster walls. The exterior is pierced with a variety of different-sized windows that cover almost 40% of the exterior facade. The windows are steel framed, and single paned, quarter-inch-thick glass with sunscreens in some areas. The windows are inset slightly for sun shading. The ceilings are lay-in acoustical tile material for all the administrative floors and lightweight sheet metal for the manufacturing and other nonadministrative working areas.

The building is designed with a main core surrounding the open courtyard at the first floor. The main movement of occupants is located at the entrance to the building and in the lobby that allows access to the manufacturing floors and to the passenger elevators near the opening of the courtyard. The courtyard core area also provides two emergency stairways that exit toward the northern vehicle parking area and driveway at the east side of the building. A utility shaft is also located adjacent to the open courtyard for electric and communication cables, security systems, and HVAC ducts.

Adversary Attack Scenario

Although the building site perimeter permits the placement of a vehicle bomb anywhere around its fence-line/curb-line, at varying standoff distances from the building, the most likely target location that would create significant damage is the southeastern corner of the building. Here, the minimal standoff distance is 25 feet, and one of the building's critical steel-concrete composite columns is fully exposed to a potential explosive attack. In practice, the blast analysis required by a risk assessment process can be performed by obtaining professional judgment from a panel of experts, by applying blast curves and engineering tables of blast loadings to this building, or by completing a computer-based finite-element analysis of the detailed structure-blast interaction. For this example, it is assumed that a vehicle carrying 500 pounds (TNT-equivalent) of bulk explosives will be placed within 25 feet of the building column and detonated before the security force can neutralize the attack.

The adversary will develop the scenario to achieve the most damage and the most destructive impact on the owners and country. Before an attack, the adversary will most likely review the drawings and examine the target of the attack for details that will ensure success and maximum destruction. Therefore, the predicted extent of structural and collateral effects and damage and the estimated human casualties for this scenario must be based on technical experience, professional background knowledge, and on actual data from experiments and explosive tests that have been conducted by numerous government and private agencies in the past. This knowledge base is readily available to those who have the need to know, and the information can be applied to most scenarios determined to be the largest risks to the owner and to the building operation. For example, extensive data is available from reports of actual attacks at U.S.-occupied buildings, including the military buildings in Beirut and Saudi Arabia; the bombing of the foundation in the parking area of the World Trade Center towers; the attacks on U.S. embassies in Africa; and the attack at the federal building in Oklahoma City. The information accumulated by our country and other countries is useful in understanding the behavior of buildings under explosive loadings and in estimating the extent of destruction and loss of lives that would result. A structural and blast engineer with the appropriate knowledge and background can provide a good approximation of the consequences that will be experienced from an explosive attack on a building.

Estimated Consequences

The ATF table indicates that 500 pounds of explosives in the trunk of a vehicle will cause lethality at up to 100 feet away. Therefore, the explosion alone is estimated to cause fatalities to more than 100 building occupants, depending on their locations inside the building at the time of the explosion. The building exterior concrete envelope will provide some protection to its occupants; however, all of the windows on the south and east side of the building will be shattered, and the debris from the broken glass and flying construction materials will impact many of the occupants who are occupying the southeast corner of the building. The airblast will penetrate the building through the window openings, and the dynamic pressures alone can be fatal or seriously damage lungs, eyes, ear drums, and much of the human body. The pressure and impulse of the air blast has also been known to push human bodies for significant distances and against interior walls. Medical problems will be experienced for many who survive, including stress to the heart and other tissues, broken limbs and fractures, and the like. An estimated 150 building occupant injuries can be expected with this scenario. The dust and fumes that would emanate from the building explosion would also cause inhalation problems to building occupants and the surrounding neighborhood and possible delayed explosions from the trapped gases in the rubble.

The southeastern corner of the structure would fail completely, and the corner would most likely collapse progressively after the critical column is first fractured and weakened and then ruptured by the vertical loads that are transferred on the column from the upper remaining structure. This progressive collapse event would destroy a major portion of the building and impact all occupants who are in the vicinity and possibly survived the air blast. An estimated 50 additional occupants will be casualties from the collapsing building. This building would be destroyed beyond repair as the interior and exterior assets will be damaged and cannot be salvaged for future repairs. The cost of a new building is estimated to be $200M, and the damage to the high-value products and raw materials is estimated to be another $50M. The loss of revenue during the 18-month period while another building is built on the same site is estimated to be another $300M, unless a temporary location can be rented in the interim period. The overall consequences are considered *High* from a 500-pound explosive attack described above, from the loss of lives and injuries, loss of revenue, loss of property value, loss of continuity of operations, and loss of credibility within the employee and customer base.

13.11   ESTIMATION OF RISK

Security risk is a function of the likelihood of attack, consequence of successful attack, and security system ineffectiveness. To estimate relative security risk, the qualitative estimates for likelihood of attack, system ineffectiveness, and consequence are logically combined. A simple method, based on expert judgment, for combining the three risk parameters to estimate security risk can be used. Table 13.19 summarizes the results of the security risk assessment.

13.11.1   Risk Summary

Medium or higher risk levels were estimated for the terrorist, criminal, gang, and insider threats for several undesired events at the building. *High* risk describes incidents that are likely to occur, have relatively *High* consequence, and against which the security system cannot adequately protect. Specifically, for the building, the threats and associated undesired events posing *Medium* or higher security risk level are:

Terrorists

- Sniper
- Bomb (exterior and interior)
- Chem/bio attack
- Cyber attack (loss of mission or theft of information)

Criminals

- Armed robbery
- Arson
- Cyber-attack (loss of mission or theft of information)

Table 13.19 Security Risk Assessment Levels

Adversary	Terrorist				Criminal				Gang				Production / Support				Business / Administration				Security Officer				System Administrator			
	P_A	$1-P_E$	C	R	P_A	$1-P_E$	C	R	P_A	$1-P_E$	C	R	TS	$1-P_E$	C	R	TS	$1-P_E$	C	R	TS	$1-P_E$	C	R	TS	$1-P_E$	C	R
Sniper	H	H	H	H																								
Bomb-exterior	H	H	H	H																								
Bomb-interior	H	H	H	H																								
Chem/bio attack	H	H	H	H																								
Arson					M	H	H	H	M	H	H	H																
Armed robbery					M	H	H	H	M	H	H	H																
Drive-by shooting									H	H	H	H																
Violent insider													H	H	H	H	H	L	H	L	H	H	H	H				
Insider sabotage eqpt.													H	H	M	M	H	L	M	L	H	H	M	M				

(continued overleaf)

Table 13.19 (continued)

Adversary	Terrorist				Criminal				Gang				Production/ Support				Business/ Administration				Security Officer				System Administrator			
	P_A	1−P_E	C	R	P_A	1−P_E	C	R	P_A	1−P_E	C	R	TS	1−P_E	C	R	TS	1−P_E	C	R	TS	1−P_E	C	R	TS	1−P_E	C	R
Insider bomb													H	H	H	H	H	L	H	L		H	H	H				
Insider theft of item													H	H	L	L												
Theft of info assets: physical	H	H	M	M	H	H	M	M					H	H	M	M	H	H	M	M	H	H	M	M		H	M	M
cyber	H	H	M	M	H	H	M	M					H	H	M	M	H	H	M	M	H	H	M	M		H	M	M
Loss of mission: physical	H	H	M	M	H	H	M	M					H	H	M	M	H	H	M	M	H	H	M	M		H	M	M
cyber	H	H	M	M	H	H	M	M					H	H	M	M	H	H	M	M	H	H	M	M		H	M	M

Gang Members

- Drive-by shooting
- Armed robbery
- Arson

Insiders

- Violent insider – production/support, security force
- Bomb – production/support, security force
- Loss of mission by cyber-attack – system administrator, production/support
- Loss of information by cyber-attack – system administrator, production/support, and business/administration

Estimated risk levels are compared to a predetermined risk threshold to decide whether further analysis is required. The threshold is determined by the analysis team and the security risk managers. The analysis team together with management decided that the threshold risk value for the corporation is *Medium*.

13.12   RISK REDUCTION STRATEGIES

If the estimated baseline risk level for the threat spectrum is judged to be above the established threshold (too *High*), risk reduction strategies for the system may be considered. Risk reduction strategies focus on reducing the levels of the parameters of the security risk equation: likelihood of attack, system ineffectiveness, and consequence. In practice, risk reduction is made most successful by improving protection system effectiveness and mitigating consequences.

Risk Reduction Upgrades – Security system planners must address how to reduce security risk. Planners might consider adding features to increase physical or cyber-protection system effectiveness and/or to reduce or mitigate consequences. Site-specific vulnerabilities identified in the system effectiveness analysis provide guidance for adding/modifying features. Upgrades to the system might include retrofits, additional safeguard features, or additional safety mitigation features. Consequence analysis and system effectiveness analysis should then be repeated for the upgraded system in order to estimate a risk level associated with the upgraded system. If the estimated risk for the upgraded system is below the threshold, the upgrade is completed. If the risk is still above the threshold, the upgrade process should be repeated until the risk level is judged to be below the threshold.

After reviewing the adversary scenarios estimated to be *High* risk to the building, an upgrade package to the physical protection system was suggested. *High* risk scenarios were estimated for the terrorist, criminal, gang, and insider threats, first. Specific upgrades were selected to increase security system effectiveness (reduce adversary success). The upgrades range from procedural changes and elimination of activities to hardware additions. Whereas general upgrades were expected to affect all scenarios, some upgrades were suggested for specific threats, such as terrorist bomb, insider, or gang drive-by shooting.

13.12.1 Physical Protection System Upgrades

A significant vulnerability in the current or base security system for the example building is the detection system. An upgrade to the system must include an upgrade to the Control Center. A perimeter intrusion detection system would allow security officers in the Control Center to receive, assess, and (video) record alarms and be able to communicate to local law enforcement. Sallyport doors into the courtyard should be closed, locked, and alarmed, and the entrance should be controlled by a personnel identification system that incorporates checking credentials and biometric identifiers. The perimeter intrusion detection system would include intrusion sensors, alarm communication, lights, and CCTV assessment capability with supervised lines and tamper protection for hardware. All exterior doors (dock, roof, emergency exits, personnel entrances) should be hardened ($\frac{1}{2}$-inch steel plate, if possible) and equipped with penetration and position sensors with CCTV assessment. In addition, the shipping/receiving doors should be hardened, locked, and alarmed and have CCTV assessment capability except during a shipment. During a shipment, a dedicated security officer in a protected structure should monitor all activity through the door. An additional protected officer should be positioned near the open vault during shipments.

Security officers would benefit from protected positions or structures, body armor, and duress signaling capability. Local law enforcement response times must be reduced. In addition, the response force must be of sufficient number to neutralize the adversary threat; arrival one at a time would not be as effective.

The physical security system upgrades are summarized below.

Control Center

- Access control-credential, biometric identifier
- Alarm communication and display upgrade

Intrusion Detection System

- Supervised lines
- Tamper-indicating devices
- Perimeter (sensors, alarm communication, lighting, CCTV assessment)
- Exterior doors including roof, shipping, emergency exits, and sallyport entrances, plus asset-control interior doors, hardened ($\frac{1}{2}$-inch steel plate added, if possible), penetration sensors, door switches, CCTV assessment, dedicated and

protected security officer positions when shipping and vault doors open

- Sensors and cameras to cover roof

Security Officers

- Protected (hardened positions)
- Duress signaling capability/communication
- Body armor

Terrorist Bomb Scenarios

- Close adjacent streets to vehicles or restrict curbside parking (may not be feasible)

Insider

- Compartmentalized work areas (close, lock, alarm, and control access to work areas by badge reader and password)
- Secure passwords
- Control of on-site items, such as tools that could be used to harm, destroy, or make explosive items
- Extensive background check, higher standard for employment, scheduled updates
- Emergency evacuation – screen for metal, evacuate to secure holding area
- All employees pass through contraband detection screening

Wastewater, Power Generators, Fresh-Air Intake

- Hardened barrier covers
- Backup sources
- Lock and alarm doors to sallyport

13.12.2   Result of Physical Protection System Upgrades

Upgrades were suggested for adversary scenarios estimated to be *High* risk for the example building. The system effectiveness values and risk values estimated for the upgrades, collectively, are given and compared to the baseline physical security system in Figure 13.16. The risk associated with some of the scenarios would be reduced with the increase in physical protection system effectiveness bought about by the implementation of features of the upgrade package.

Adversary:	Terrorist	Criminal	Gang	Insider		
				Production /support	*Business /administration*	*Security Officer*
Undesired event:						
Sniper	Low|Low					
Bomb-exterior	Low|Low					
Bomb-interior	Low|High					
Chem./Bio. attack	Low|High					
Arson		Low|High	Low|High			
Physical attack –cyber eqpt.	Low|High	Low|High				
Armed robbery		Low|High	Low|High			
Drive-by shooting			Low|Low			
Insider - violence				Low|Low	High	Low|Low
Insider - sab. eqpt				Low|Low	High	Low|Low
Insider - bomb				Low|High	High	Low|High
Insider - theft of item				Low|High	N/A	N/A

Condition 1|Condition 2

Figure 13.16   Physical Protection System Effectiveness for Upgraded Package vs. Baseline System.

Exterior Bomb

Closing adjacent streets is deemed to be infeasible in the case for the example building. The building is still vulnerable to a vehicle bomb attack on the north side. No change in physical protection system effectiveness or risk level is expected.

Sniper, Drive-By Shooting

Inevitably, even with the provisions of the upgrade package, there are still times that employees/personnel will be outside the building structure and susceptible to a sniper or drive-by shooting incident. No change in physical protection system effectiveness or risk level is expected.

Armed Robbery

Condition 1 – for this scenario, the adversary climbs or bridges the fence, traverses the area to the shipping/receiving door that is open for shipments and manned by a security officer, enters the open building, travels to open vault, packs up finished product assets, and exits the same way as entrance. Condition 2 – same scenario except building doors and vault doors are closed and locked. Figure 13.17 summarizes the estimation of protection system effectiveness for the upgraded system. System effectiveness was judged to be *Low* for the baseline system, *Low* for Condition 1, and *High* for Condition 2 if law enforcement can respond within 180 seconds.

Interior Bomb, Arson, Chemical / Biological Attack in the Courtyard

Condition 1 – for this scenario, the adversary climbs or bridges the fence, traverses the area and enters via the sallyport doors, enters the courtyard, sets explosives, and retreats. Condition 2 – same scenario. Figure 13.18 summarizes the estimation of protection

System Features	FEN2	PA2		SHD1(C)		Bldg Int			DOR	Task
Detection Effectiveness	H			H					H	H
Delay (s.) Cond 1|2	10	7(500̃)		0	30	10(700̃)	0	180		120
A	Detection Effectiveness value (maximum Value)					High				
B	Response Effectiveness Value					High				
C	Sum of delays (including and after first H or second M for detection) Condition 1/2					147 s.			357 s.	
D	Response force time					180 s.				

E	Compare C to D:
	  L if C < D
	  M if C ~ D
	  H if C > D
						C < D			C > D	

	SUMMARY OF SYSTEM EFFECTIVENESS (minimum of A, B, and E)					Cond. 1 - Low			Cond. 2 - High	

Figure 13.17   Scenario Analysis for Armed Robbery.

System Features	FEN(A)	PA 2		PER	Courtyard		Task
Detection Effectiveness	H			H			H
Delay (s.) Cond 1&2*	10	7(500̃)		30	15(1000̃)		180
A	Detection Effectiveness Value (maximum Value)				High		
B	Response Effectiveness Value				High		
C	Sum of delays (including and after first H or second M for detection) Condition1 & 2*				242 s.		
D	Response force time				180 s.		

E	Compare C to D:
	  L if C < D
	  M if C ~ D
	  H if C > D
					C > D		

	SUMMARY OF SYSTEM EFFECTIVENESS (minimum of A, B, and E)				Both Conditions - High		

Figure 13.18   Scenario Analysis for Attacks in Courtyard.

system effectiveness for the upgraded system. System effectiveness was judged to be *Low* for the baseline system, and *High* for both Conditions 1 and 2 if law enforcement can respond within 180 seconds.

System Features	FEN2	PA 2	SHD1 (C)		Bldg Int	CC DOR	Task
Detection Effectiveness	H		H			H	H
Delay (s.) Cond 1/2	10	7(50')	0	30	20(150')	30	180
A	Detection Effectiveness Value (maximum Value)				High		
B	Response Effectiveness Value				High		
C	Sum of delays (including and after first H or second M for detection) Condition 1/2				247 s.		277 s.
D	Response force time				180 s.		
E	Compare C to D: L if C < D M if C ~ D H if C > D				C > D		C > D
	SUMMARY OF SYSTEM EFFECTIVENESS (minimum of A, B, and E)				Cond. 1-High		Cond. 2- High

Figure 13.19   Scenario Analysis for Control Center Attack.

Information System Physical Attack at Control Center

Condition 1 – for this scenario, the adversary climbs or bridges the fence, traverses the area and enters the building via open shipment doors, travels to the Control Center, defeats the door, enters the Control Center, sets explosives, and retreats. Condition 2 – same scenario, except that the building doors are closed and locked. Figure 13.19 summarizes the estimation of protection system effectiveness for the upgraded system. System effectiveness was judged to be *Low* for the baseline system; and *High* for both Conditions 1 and 2 if law enforcement can respond within 180 seconds.

Insider – Violence and Sabotage of Equipment

Scenario A or B (insider adversary attempts to carry in contraband or insider adversary attempts to bring in the contraband via alternate routes) – the upgrade procedure that requires all personnel to pass through contraband detection process together with screening all incoming materials results in *High* protection system effectiveness for all insider categories. For Scenario C (insider adversary acquires or constructs weapon, flammable, or bomb inside the production area) – training personnel to detect unusual behavior and the elimination of materials and equipment that could be made into a weapon is not expected to significantly increase protection system effectiveness.

Insider – Bomb or Theft of Item

Because of the upgrade package requirement for all personnel to be screened for contraband on entry and exit, the physical protection system effectiveness is judged to be *High* for preventing the insider from bringing in a bomb or removing a valuable product.

Conclusion

Because physical protection system upgrades cannot prevent all identified undesired events from occurring except in cases in which law enforcement must arrive in such a short time that they almost have to be located on-site, the upgrade package should also include consequence mitigation schemes in order to reduce the security risk.

13.12.3   Cyber-Protection System Upgrades

Authentication

- Implement strong passwords
- Add two-factor authentication

Authorization

- Compartmentalize authorized access to business IT network and production process control network
- Compartmentalize authorized access for computer system administrators

Audit

- More frequent review of traffic data

Integration

- Add firewalls to business system and process control system electronic security perimeters
- Encrypt all communications into and within business system and process control system

13.12.4   Results of Cyber-Protection System Upgrades

The addition of the cyber-protection system upgrades is expected to significantly enhance the effectiveness to the *High* level.

Loss of Mission / Theft of Information

Outsider threat – The additions of stronger authentication, authorization, audit, and integration features are expected to make it significantly more difficult to cause loss of mission or impact the business IT system to cause loss of proprietary information.

Insider threat – The addition of stronger authentication, authorization, audit, and integration features are expected provide better protection against the system administrator position to be ability to cause loss of mission or impact the business IT system to cause loss of proprietary information.

13.12.5   Consequence Mitigation Upgrades

One of the most extensive and costly consequence mitigation upgrades that has been applied to buildings is for structural hardening against a bulk explosive charge delivered in a vehicle by a terrorist. If this scenario is in fact deemed to have a *High* likelihood of attack at a particular building, the most effective measure to apply is the additional standoff distance to the building's key structural members, such as a column. This is usually possible if the building is situated in a rural area and there is sufficient acreage to allow for a vehicle barrier system at 100 to 200 feet away from the building. This additional standoff distance is usually achieved by using vehicle-arresting cable attached to rigid post at the building site perimeter; installing bollards, planters, and/or rigid walls to arrest vehicles, or by increasing the height of the curb sufficiently. Each of these barriers in its own way will act as an anti-ram device and prevent adversarial vehicles from getting any closer to the building. The effects of an explosive charge attack are significantly diminished with distance away from a designated target.

The minimal distance from the curb to the building is 25 feet in this example, and there are no means currently for preventing a vehicle with explosives from ramming through the fence and traveling up against the building columns at the south corners. Therefore, one of the first hardening measures is to ensure the maximum standoff distance is maintained. The streets around the east and west sides of the building must be closed to any vehicular traffic. The south side has 50 feet of set-back (standoff), but this distance is insufficient to protect the building from a 500-pound charge. In interest of maintaining these set-backs from the fence, a vehicle-arresting cable well mounted to the fence posts will assure that a large amount of bulk explosives cannot be placed near the building. The north side of the building has 150 feet of set-back that extends to the site perimeter of the parking area. This set-back appreciably reduces the effects of the 500-pound charge; windows will shatter and some damage to the façade will be noticed. However, installing the arresting cable along this site perimeter will be of benefit by providing an assured ram-free boundary along the entire fence line. To further protect the building from the effects of blast, soil-rock berms can be constructed within the fenced area and the building itself. These soil berms up to 6 feet high will act as backup vehicular barriers and blast mitigation at the lower floors. The cost for the arresting cable installation on the existing fence, including along any of the gates to the site, and building soil berms with the property site, is estimated to be $0.5M.

The other consequence mitigation upgrade to be considered is hardening the structure so that progressive collapse of the building is not expected. The most common means for structural hardening is wrapping a carbon fiber embedded in resin plastic around the column and on the wall surfaces up to about the third or fourth floors from the ground. In addition, the columns will need to be reinforced with cross braces to the floor beams up to about the fourth or fifth floors, on all corners of the building, to allow for load transfer in the event of a column collapse nearest to the explosive charge. These techniques must be evaluated using blast-structural analysis methods to verify that potential collapse will be averted and that the damage to the remaining structural systems is considered minimal with these upgrades. The cost of these structural hardening measures is estimated at $5M, including the costs of some operational downtime during the construction of these upgrades.

The windows on the first four exposed floors would also require some hardening measures, such as film application to minimize chards and fragments from impacting the occupants. This upgrade technique is used; however, the security enhancement is still questionable because of the potential for the entire glass pane to act as a projectile that impacts building occupants. Alternatively, the glazing on the first four exposed floors can also be replaced with a more blast-resistant glass, such as a laminated glass, and adding anchor-frame-mullion upgrades, where needed. The estimated cost for replacement glazing is about $1.5M.

13.12.6 Summary

These consequence mitigation upgrades will reduce the risk from *High* to *Medium*. The loss of life because the building will not collapse is reduced from 150 to less than 50, and injuries from falling objects and flying glass and debris will be reduced from 150 to less than 50. The cost for restoring the building and its interior equipment will be reduced from $200M replacement cost to $50M repair costs. The loss of high-value products and inventories will be reduced from $50M to $5M primarily because the vault will remain intact. The loss of revenue will be reduced from $300M to $50M because the repair time will be less than 6 months compared with 18 months for total building replacement. The total cost for consequence mitigation was estimated at $7M. The savings is projected to be $445M ($150M, $45M, $250M for building repairs, loss of high-value inventory, and loss of revenue, respectively). One hundred building occupant lives saved and 100 reduced occupant injuries indicate the magnitude of risk reduction possible from the building hardening measures described.

These risk reduction factors do not include the benefits that would be derived from the enhanced public and customer confidence and occupant comfort from knowing that upgrade measures would provide added protection in the event of a deliberate explosive attack. The insurance premiums might be reduced in the longer term if the upgrades are evaluated through appropriate negotiations regarding the potential for reduced liabilities and lawsuits. The neighboring community might believe in the reduction in target attractiveness and thereby consider itself much safer without the burden of the looming threat. The cost of the upgrades could be amortized in a selected number of years and the return on the investment might then be appreciably higher.

13.13   IMPACT ANALYSIS

Impact Analysis – Once the system upgrade has been determined, it is important to evaluate the impacts of the risk reduction on the mission of the facility and the cost. If system upgrades put a heavy burden on normal operation, a trade-off would have to be considered between risk and operations. Budget can be the driver in implementing security upgrades. A trade-off between risk and total cost may have to be considered. The assessed level of risk and the upgrade impact on cost, mission, and schedule is valuable information to security risk managers. Figure 13.20 describes the expected impacts of implementing the proposed upgrade package.

13.13.1   Impacts of Upgrade Package

The upgrade package is assessed to have:

- No negative impact on security risk level; in fact, security risk is reduced by the upgrade package. See Table 13.20.

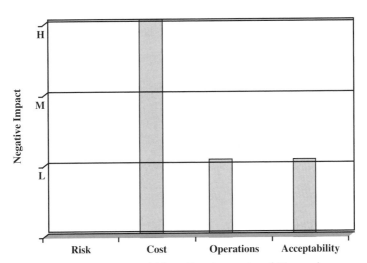

Figure 13.20   Expected Negative Impacts of Upgrade Package.

Table 13.20 Comparison of Baseline Security Risk and Security Risk Reduction Afforded by the Upgrade Package

Adversary	Terrorist				Criminal				Gang				Production/ Support				Business/ Administration				Security Officer				System Administrator			
	P_A	$1-P_E$	C	R	P_A	$1-P_E$	C	R	P_A	$1-P_E$	C	R	TS	$1-P_E$	C	R	TS	$1-P_E$	C	R	TS	$1-P_E$	C	R	TS	$1-P_E$	C	R
Sniper	H	H|H	H	H|H																								
Bomb-exterior	H	H|H	H|M	H|M																								
Bomb-interior	H	H|L	H|M	H|M																								
Chem/bio attack	H	H|L	H	H|L																								
Arson					M	H|L	H	H|L	M	H|L	H	H|L																
Armed robbery					M	H|L	H	H|L	M	H|L	H	H|L																
Drive-by shooting									H	H|H	H	H|H																
Violent insider													H	H|H	H	H|H	H	L	H	L	H	H|H	H	H|H				
Insider sabotage eqpt.													H	H|H	M	M|M	H	L	M	L	H	H|H	M	M|M				

(continued overleaf)

Table 13.20 (continued)

Adversary	Terrorist				Criminal				Gang				Production/ Support				Business/ Administration				Security Officer				System Administrator			
	P_A	$1-P_E$	C	R	P_A	$1-P_E$	C	R	P_A	$1-P_E$	C	R	TS	$1-P_E$	C	R	TS	$1-P_E$	C	R	TS	$1-P_E$	C	R	TS	$1-P_E$	C	R
Insider bomb													H	H|L	H	H|L	H	L	H	L	H	H|L	H	H|L				
Insider theft of item													H	H|L	L	L|L												
Theft of info. assets: physical	H	H|L	M	M|L	H	H|L	M	M|L					H	H|L	M	M|L	H	H|L	M	M|L	H	H|L	M	M|L	H	H|M	M	M|M
cyber	H	H|L	M	M|L	H	H|L	M	M|L					H	H|L	M	M|L	H	H|L	M	M|L	H	H|L	M	M|L	H	H|M	M	M|M
Loss of mission: physical	H	H|L	M	M|L	H	H|L	M	M|L					H	H|L	M	M|L	H	H|L	M	M|L	H	H|L	M	M|L	H	H|M	M	M|M
cyber	H	H|L	M	M|L	H	H|L	M	M|L					H	H|L	M	M|L	H	H|L	M	M|L	H	H|L	M	M|L	H	H|M	M	M|M

- High negative impact on cost. Addition of physical, cyber-, and blast effects protection magnify construction, maintenance, and operation costs.
- Low negative impact on operations and acceptability by personnel. Upgrade package is not expected to impact day-to-day activities and so is expected to be accepted by personnel.

13.13.2   Impacts of Consequence Mitigation Package

The most significant impact for these blast consequence mitigation measures would be the investment cost for the approved upgrades. There would also be some impact on the security operations regarding the arresting cable at the entry gates, and periodic inspections required along the fence-line that would better ensure adequate anti-ram performance throughout its life expectancy. The soil berms adjacent to the building might obscure the visibility from inside the building basement. The impact of construction modifications to the occupants and to the operations would be noticeable but minimal because most of the work will be on the main building structure and specifically along the exterior walls.

13.14   PRESENTATION TO MANAGEMENT

The final step in the risk assessment process is the preparation of a presentation package for the risk managers and stakeholders. This material is usually presented in a briefing, which affords decision makers the opportunity to request clarification or additional data and ask any questions about the assessment, including its assumptions, conclusions, recommendations, and supporting data.

The presentation generally includes the threat description, the security risk estimates for the baseline system, descriptions of any risk reduction packages, and the results of the impact analysis for the risk reduction package(s). By comparison to the baseline risk levels, managers are able to understand what the upgrade package is buying them in risk reduction as well as other potential impacts. The total presentation package provides invaluable information for risk management decision makers.

13.14.1   Threat Description

The threat for this building is presented in Tables 13.3 through E-8, the summary tables that show the threat information and analyses performed for the building. Depending on the level of detail required by the management and stakeholders for whom the presentation is prepared, this may be adequate. Table 13.3 lists the five types of adversaries considered for the building – terrorist outsiders who may be colluding with an insider, an insider, criminals, extremists, and a gang, and describes the numbers of individuals associated with each group; the equipment, vehicles, and weapons available to each group; and the tactics employed by the group. These characteristics serve to define and distinguish the potential adversaries. Tables 13.4 through E-8 specify the likelihood of attack by each of these five types of adversaries by matching such threat characteristics as capability, historical and current interests, previous attacks, current surveillance and documented threats, consequences, and ideology with specific actions that would cause undesired events. The insider threat is approached somewhat differently, as the insider security threat is defined by access to assets, authority, access to the security system, and the opportunity for collusion.

13.14.2   Security Risk Estimates for the Baseline System

Security risk assessment levels (based on the consequences associated with a specific attack, the likelihood of such an attack by a potential adversary, and the likelihood of the physical protection system at the building failing to prevent this attack) for the building security events considered in this risk assessment are summarized in Table 13.18. Again, the insider threat is considered somewhat differently, as the severity of the threat is variable for insider threats (depending on access, authority, and opportunity) and thus the security risk is variable. Table 13.18 shows the security risk variability of the insider threat by summarizing different positions.

13.14.3   Risk Reduction Packages

For the baseline security risk levels considered too *High* – that is, above the specified threshold established for this risk assessment – a risk reduction package was prepared. Risk reduction strategies focus on reducing one or more parameter of the security risk equation: likelihood of attack, security system ineffectiveness against that attack, and the consequences of a successful attack.

13.14.3.1   Reducing the Likelihood of Attack

It is unlikely the building would be able to affect the many characteristics of the adversary, with the possible exception of ease of attack. Even if the building security decision makers were able to reduce the ease of attack, how would it be measured? And would this reduction be sufficient to reduce the overall security risk? In practice, security risk reduction is very difficult to achieve by reducing the likelihood of attack.

13.14.3.2   Reducing Security System Ineffectiveness

Upgrade packages for the physical protection system and the cyber-protection system were developed to reduce the security risk of the building. The physical security system upgrades are summarized below:

Control Center

- Access control – credential, biometric identifier
- Alarm communication and display upgrade

Intrusion Detection System

- Supervised lines
- Tamper-indicating devices
- Perimeter (sensors, alarm communication, lighting, CCTV assessment)
- Exterior doors, including roof, shipping, emergency exits, and sallyport entrances, plus asset-control interior doors, hardened ($\frac{1}{2}$-inch steel plate added, if possible), penetration sensors, door switches, CCTV assessment, dedicated and protected security officer positions when shipping and vault doors open
- Sensors and cameras to cover roof

Security Officers

- Protected (hardened positions)
- Duress signaling capability/communication
- Body armor

Terrorist Bomb Scenarios

- Close adjacent streets to vehicles or restrict curbside parking (may be infeasible)

Insider

- Compartmentalized work areas (close, lock, alarm, and control access to work areas by badge reader and password)
- Secure passwords

- Control of on-site items, such as tools that could be used to harm, destroy, or make explosive items
- Extensive background check, higher standard for employment, scheduled updates
- Emergency evacuation – screen for metal, evacuate to secure holding area
- All employee pass through contraband detection screening

Wastewater, Power Generators, Fresh-Air Intake

- Hardened barrier covers
- Backup sources
- Lock and alarm doors to sallyport

In addition to the bricks-and-mortar security upgrades, the protection of the data and proprietary information in the building's cyber-system should also be improved. Upgrades to the cyber-protection system include:

Authentication

- Implement strong passwords
- Add two-factor authentication

Authorization

- Compartmentalize authorized access to business IT network and production process control network
- Compartmentalize authorized access for computer system administrators

Audit

- More frequent review of traffic data

Integration

- Add firewalls to business system and process control system electronic security perimeters
- Encrypt all communications into and within business system and process control system

13.14.3.3   Mitigating the Consequences of a Successful Attack

Consequence mitigation is another very effective way to reduce security risk. Increasing the standoff distance and hardening the structure can significantly reduce the consequences of a bulk explosive charge delivered in a vehicle by a terrorist.

The first recommended mitigation measure is to create the maximum standoff distance. Significant protection would be provided if the east and west sides of the building could be closed to any vehicular traffic. The south side has 50 feet of set-back (standoff), insufficient to protect the building from a 500-pound charge. A vehicle-arresting cable well-mounted to the fence posts will assure that a large amount of bulk explosives cannot be placed near the building. The north side of the building has 150 feet of set-back that extends to the site perimeter of the parking area. This set-back appreciably reduces the effects of the 500-pound charge; still, installing the arresting cable along the entire site perimeter will provide an assured ram-free boundary along the fence-line. To further protect the building from the effects of blast, soil-rock berms can be constructed within the fenced area. These soil berms, up to six feet high, act as backup vehicular barriers and provide blast mitigation to the lower floors.

The next consequence mitigation recommendation is hardening the structure so that the progressive collapse of the building is not expected. Wrapping the columns with a carbon fiber embedded in resin plastic and applying this material to the wall surfaces up to about the third or fourth floors will dramatically diminish the likelihood of collapse. The columns should be reinforced with cross braces to the floor beams up to about the fourth or fifth floors, on all corners of the building, to allow for load transfer.

The windows on the first four exposed floors also require hardening measures. The recommendation is to replace the existing glass with a more blast-resistant glass, such as a laminated glass, and to include anchor-frame-mullion upgrades where needed.

These consequence mitigation upgrades will reduce the consequence level associated with a vehicle bomb from *High* to *Medium*. The loss of life because the building will not collapse is reduced from 150 to less than 50. Injuries from falling objects and flying glass and debris also will be reduced from 150 to less than 50. The cost for restoring the building and its interior equipment will be reduced from $200M replacement cost to $50M repair costs.

13.14.4 Impact Analysis for Risk Reduction Package

The upgrade package is assessed to have:

- No negative impact on security risk level; in fact, security risk is reduced by the upgrade package.
- High negative impact on cost. The cost for the arresting cable installation on the existing fence and gates to the site and building soil berms is estimated to be $0.5M. The cost of wrapping columns and walls to harden the structure is estimated at $5M. The estimated cost for replacement glazing is about $1.5M. Addition of physical, cyber-, and blast effects protection magnify construction, maintenance, and operation costs.
- Low negative impact on operations and acceptability by personnel. The upgrade package is not expected to have much impact on day-to-day activities and so is expected to be accepted by personnel.

13.15   RISK MANAGEMENT DECISIONS

Building owners, stakeholders, and security risk managers have the risk assessment information package to help them make difficult security decisions. Many options are available. The purpose of the security risk assessment is to provide the decision makers with the data and analyses required to make an informed decision. Risk managers can decide to:

- Accept the security risk level of the baseline system, if the consequences do not exceed the threshold of acceptability. A risk manager might select this option when the consequences of an attack or undesired event are less costly in some way than preventing the attack or mitigating the result. In this case, 150 deaths are unacceptable.
- Buy more insurance to offset the costs of *High* consequences should an adversary attack successfully. If the consequences are less than devastating, this could be a cheaper way to manage risk. However, in this case, the consequence of 150 deaths is devastating.
- Implement the recommended risk reduction packages. Risk can be reduced by increasing protection system effectiveness and/or by mitigating consequences. Consequence mitigation usually involves people, procedures, policies, training, and equipment. Consequence mitigation is an appealing choice for a building or facility because generally it is a more affordable approach for reducing risks than buying physical protection technologies. In this case, the risk managers opted to implement a combination of consequence mitigation and improved protection system effectiveness.
- Ask the analysis team for additional analyses or clarification. Management did request clarification of the team on a number of issues. The analysts were able to satisfy the risk

managers' information needs using the already collected data and existing analyses.

- Provide contingency measures for security risks that cannot be covered at all times, but can be implemented during periods of heightened threat conditions. Management will improve evacuation plans and develop responses to alerts that will minimize injuries and deaths. For example, if adjacent streets cannot be permanently closed to traffic, they could be closed during *High* threat conditions.

- Establish a threat-level description that describes a subset of the site-specific threat that can be protected against right now, with plans for addressing the higher-level threats as resources permit. In this case, the upgrades to the physical protection system can be accomplished more quickly than the consequence mitigation features of installing the perimeter cables, adding berms, wrapping the columns and walls, and reglazing the windows.

While a formal decision has not yet been reached, the managers are discussing upgrading the physical protection and the cyber-protections systems immediately, developing contingency plans and training the employees in following them in the short term, and budgeting for the incremental hardening of the facility over the next three fiscal years.

Deciding the appropriate response to an identified risk is the province of risk managers. The key to a successful decision is in knowingly determining a risk level that is acceptable, given the available resources, rather than unwitting acceptance of an existing amorphous risk. The purpose of the security risk assessment is to provide the decision makers with the information required to make successful decisions.

Appendix A

Generic Fault Tree for Buildings

NOTES ON FAULT TREES

Fault Tree Analysis – An analytical technique, whereby an undesired state of the system is specified (usually a state that is critical from a safety or security standpoint), and the system is then analyzed in the context of its environment and operation to find all credible ways in which the undesired event can occur.

Fault Tree – A diagram that graphically and logically depicts the interrelationships of elementary events that lead to an undesired event (called the "top event" of the fault tree).

Completeness – key to building a good fault tree is *completeness*. Completeness is achieved by:

- Careful definition of each event (fault)
- Taking *small* steps in logic
- Being exhaustive at each step
- Considering faults within this "subsystem"
- Considering faults in other subsystems that supply or otherwise support this subsystem
- Considering events external to the system that can cause faults in this subsystem

- Being absolutely sure that your definition of subsystems does not leave out *any* part of the system – no matter how trivial.

Top Event – The *single* event (usually undesired), or fault, for which the list of potential causes is sought.

Intermediate Events – A fault that represents a state that contributes to the top event but is not itself a failure. This event occurs as a result of one or more antecedent causes, and requires further development. Intermediate events are represented in fault trees by "AND" and "OR" gates (or other logical structures).

Primary Event – A *failure* that represents an *elementary cause* of the preceding intermediate event. These events require no further development because:

- Further development may not be possible.
- Further development may be beyond the scope of the analysis.
- Types of primary events:
    - **Basic Event** – An elementary cause of component failure.
    - **External Event** – An external agent acting on the system causes a failure within the system.
    - **Undeveloped Event** – Further development not undertaken.
    - **Developed Event** – Further development of this event occurs in some other model – possibly another fault tree model.

SYMBOLS

Primary Event Symbols:

 Basic Event (BE) – An initiating fault requiring no further development

Conditioning Event (CE) – Specific conditions or restrictions that apply to any logic gate (used with PRIORITY AND gates)

Undeveloped Event (UE) – An event that is not further developed either because it is of insufficient consequence or because information is unavailable

Developed Event (DE) – An event that could be further developed or is developed elsewhere but is treated here as a primary event

External Event (EE) – An event that is normally expected to occur (usually an event having probability of one or zero)

Gate Symbols:

"AND" Gate (AG) – Output fault occurs if all of the input faults occur

"OR" Gate (OG) – Output fault occurs if at least one of the input faults occurs

"Exclusive OR" Gate (EOG) – Output fault occurs if exactly one of the input faults occurs

"Priority AND" Gate (PAG) – Output fault occurs if all of the input faults occur in a specific sequence (the sequence is represented by a CONDITIONING EVENT drawn to the right of the gate)

"Special" Gate (SG) – Output fault occurs according to a logic function defined by the user

Miscellaneous Symbols:

Description – Contains the description of an event

Transfer In – Indicates that the tree is developed further at the occurrence of the corresponding "Transfer Out" (usually on another page)

Transfer Out – Indicates that this portion of the tree must be attached at the corresponding "Transfer In"

Generic Fault Tree for Buildings:

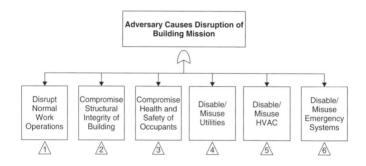

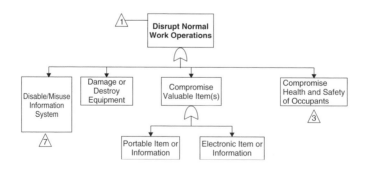

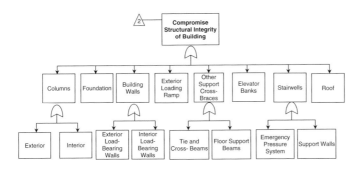

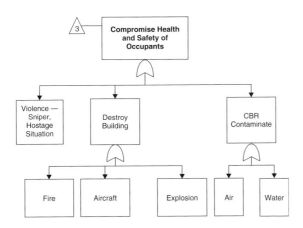

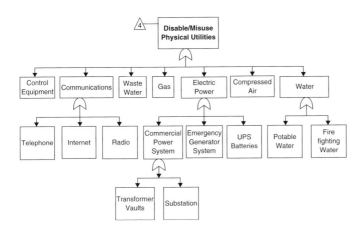

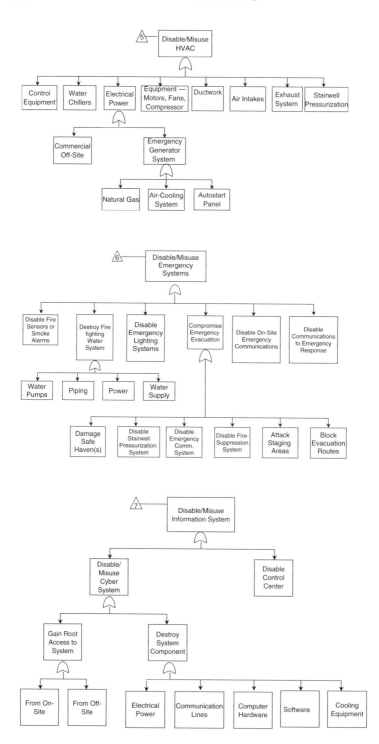

Appendix B

Adversary Sequence Diagrams

An ASD is constructed for each critical asset included in the most-vulnerable strategy. The ASDs are used to model adversary paths to the critical asset, to derive the most-vulnerable adversary scenarios, and later, to support the reduce risk (system upgrade) function.

The ASD models the physical protection system at a facility. It identifies paths that adversaries can follow to accomplish the undesired event. An ASD can be used to model all possible adversary paths through a facility. ASDs for buildings may only have one or two layers of protection, but they are helpful tools. They help prevent overlooking possible adversary paths and, when considering protection system upgrades, ASDs help select upgrades that affect the most adversary paths and can help to ensure that all adversary paths are addressed. For an example, suppose that the undesired event is to interrupt or disrupt the information system by attacking the control system operations. Figure B-1 shows a sample building with two representative physical paths that adversaries might take to damage or sabotage the controls (critical asset) inside the control room. Cyber-attack scenarios will be addressed in the next section.

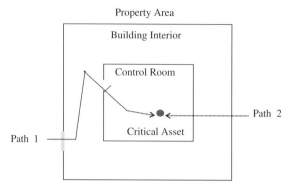

Figure B.1   Possible Adversary Paths.

There are three basic steps in creating an ASD for a specific building. These include:

1. Model the facility by separating it into adjacent physical areas.
2. Define the system features between the adjacent areas.
3. Construct the ASD.

PHYSICAL AREAS

The ASD models a facility by separating it into adjacent physical areas. Figure B-2 is a facility sketch of the example building.

Figure B-3 describes the adjacent physical areas of the example building. The ASD represents areas by rectangles.

PATH ELEMENTS AND PROTECTION SYSTEM FEATURES

The ASD models a physical protection system by identifying protection layers between the adjacent areas (see Figure B-4).

Each protection layer consists of a number of system features. The types of system features used in an ASD include:

- DOOR – Doorway

- DUCT – Duct
- FENCE – Fence line
- GATE – Gateway (could be pedestrian or vehicle)
- TASK – Task at critical asset

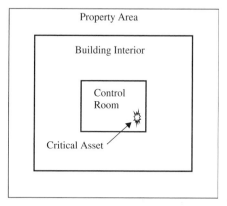

Figure B.2   Basic Areas at the Example Building.

| Off-Site |
| Property Area |
| Building Interior |
| Control Room |
| Critical Asset |

Figure B.3   Adjacent Physical Areas for the Example Building.

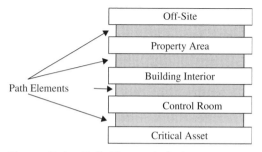

Figure B.4   Path Elements between Adjacent Areas.

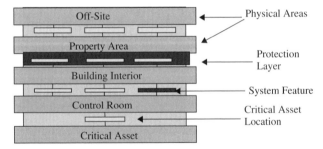

Figure B.5   ASD Concept.

- PORTAL – Series of two barriers with area between (could be gates or doors)
- SURFACE – Could be wall, roof, floor
- TUNNEL
- WINDOW

The basic ASD as it has been developed so far is given in Figure B-5. The adversary attempts to sequentially defeat a feature in each protection layer as he traverses a path through the facility to the critical asset. The ASD represents all of the realistic paths that an adversary might take to reach a critical asset. For sabotage analysis, only the entry paths would be evaluated, and the system features would be assumed to be traversed in only one direction. For theft analysis, the ASD shown should be considered to be traversed twice – on entry to the critical asset and on exit from the critical asset.

Sometimes it will be necessary to deviate from the orderly sequence of physical areas and protection layers of the generic ASD in order to create an accurate site-specific ASD. A jump is used to model a system feature that does not directly connect to the adjacent area.

Assume, for example, the facility shown in Figure B-6. There is a wall common to the building and to the critical asset enclosure. This situation is correctly modeled by including a surface jump feature from the control building to model this portion of the

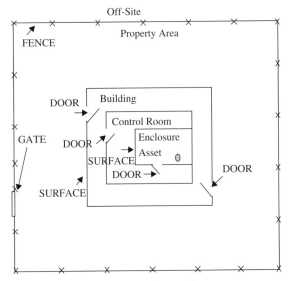

Figure B.6   Sample Facility with Jump.

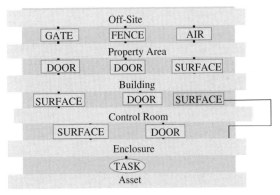

Figure B.7   ASD for Sample Facility with Jump.

common surface. As shown in Figure B-7, the ASD then shows a direct path that jumps from the building to the critical asset enclosure (without passing through the control room) in addition to all other selected indirect paths.

Appendix C

Physical System Effectiveness Worksheets

ESTIMATING PHYSICAL PROTECTION SYSTEM EFFECTIVENESS FOR COMPLEX PROTECTION SYSTEMS

The worksheets in this appendix should be used to estimate detection and delay values for path elements of a complex physical protection system.

1. List the path elements associated with the most-vulnerable scenario.
2. Review factors that could contribute to effectiveness in Table C.1.
3. Estimate the Detection function effectiveness by completing the appropriate system feature sheets. Use VL for very low, L for low, M for medium, H for high, and VH for very high. Record the estimates in the second row of the table given above. The Detection function includes:
   a. Access control (ID and authorization checks, assessment, and communication of alarm)

Table C.1 Effectiveness Factors

Hardware	Personnel	Operational Procedures
Site Conditions (terrain, vegetation, wildlife, etc.)	**Training**	**Appropriateness**
Environmental Conditions	**Personnel Background Check**	Development of contingency plans
Natural (nighttime, inclement weather, etc.)	**Motivation**	Coordination with local law enforcement
Man-made (electromagnetic interference, operational noise, etc.)	**Physical Fitness of Personnel**	Coordination with fire/rescue services
	Alertness	
	Skills	**Review and Revision**
Performance Conditions	**Dedication**	Security planning
Installation	**Ability to Perform Mission**	Changes in operations
Operation	Coordination with other security personnel (both on-site and off-site)	**Human Interface**
Maintenance		**Clarity**
Testing		**Training**
Tamper proofing	Appropriateness of assignment	Operational
Specific Vulnerabilities		Security
Compensatory Measures	Environmental conditions	**Frequency of Use**
Inoperable		**Rigorous Implementation**
Insensitive		
High false alarm rate		

    b. Contraband detection (sensors for contraband, searches, assessment, and communication of alarm)

    c. Personnel (security personnel, general observation)

    d. Intrusion detection system (sensors, assessment, and communication of alarm)

4. Estimate the delay times associated with each system feature by completing the appropriate system feature sheets. Delay times will be given in seconds. Record the delay times in the third and final row of the table for each system feature. The Delay function includes time delays (assuming detection has occurred) for:

    a. Barriers (doors, surfaces, windows, impediments . . .)

    b. Locks

    c. Security personnel

    d. Task at critical asset to cause undesired event (set up explosives, damage, . . .)

5. Estimate traversal times for the areas between the respective system features. Table C.2 provides estimates of traversal times based on modes of transportation. The times are given in seconds and should be listed on the third and final row of the table for each area.

Table C.2 Worksheet for Area Traversal Time

Estmating Delay Times

Mode	Distance in Feet	Rate	Delay Time (s) (Distance Divided by Rate)
Walking		7 ft / s	
Running		15 ft / s	
Crawling		4 ft / s	
Climbing (up or down)		1 ft / s	
Driving (pick up)		54 ft / s	

Table C.3 Response Effectiveness

Communications System with Response Force Provides for	L^a	M^b	H^c	Justifications / - Comments
Radio communications during all weather conditions				
Procedures and means to communicate with local law enforcement				
Rigorous procedures for calling off-site response				
Ability to have continuous communication during an incident related to response force posture (i.e., tactics, locations, etc.)				
Ability of control center to communicate with off-site response by wire and radio				
On-site backup power for radio communications				
Ability to authenticate communications messages				
Response force:				
Has adequate firepower to delay adversaries until backup response arrives				
Has sufficient firepower readily available to all response force members				
Has the capability to covertly signal duress to control center				
Has protected fighting positions for on-site response force				

Table C.3   (*continued*)

Communications System with Response Force Provides for	L^a	M^b	H^c	Justifications / Comments
Is large enough in numbers to neutralize the threat				
Summary				
Minimum Value of Response Force Effectiveness				

[a] *Low effectiveness:* Little evidence of capability. Some effectiveness factors not met at all. Some adversary attributes cause severe degradation of performance. Little test data available to validate performance estimates.

[b] *Medium effectiveness:* Evidence of general, but not specific, capability. Some effectiveness factors not rigorously met. Some adversary attributes cause important, but not complete, degradation of performance.

[c] *High effectiveness:* Evidence of specific capability of security system to address specific concern of questions. All effectiveness factors are rigorously met. All adversary attributes are considered in context of defined threat. Test data or a specific vendor data validates performance estimates.

6. Estimate the Response function. Table C.3 provides questions to help evaluate the effectiveness of a response by local or federal law enforcement officers. Effectiveness estimates should be recorded in the summary table below. The estimate for the law-enforcement-type response should be listed in the summary table below.

7. Estimate system effectiveness, P_E, the ability of the protection system to prevent the undesired event. In general, the questions are: Does the system have effective detection, delay, and response functions, and does the system detect early enough and have enough delay time for the undesired event to be interrupted and prevented? The table below summarizes this information.

- The detection entry is the maximum of the values of detection effectiveness across the top row of the table above.
- The response value is the summary time value for law enforcement.
- The delay entry is the delay time that occurs after effective detection takes place.
- The delay time will be counted after the first H or M value (which ever occurs first in the scenario) for detection.

A comparison is made:

a. If the delay time is less than the response time, P_E is low, L.
b. If the delay time is longer than, but close to the response time, P_E is medium, M.
c. If the delay time is comfortably longer than the response time, P_E is high, H.
d. The estimate of effectiveness of the safety/mitigation features is summarized last.

A	Detection Effectiveness Value (maximum value)	
B	Response Effectiveness Value	
C	Sum of Delays (including and after first H or M)	s.
D	Response Force Time	s.
E	Compare C to D: L if C < D M if C ~ D H if C > D	
	SUMMARY OF SYSTEM EFFECTIVENESS (minimum of A, B, and E)	

SYSTEM FEATURES FOR BUILDINGS

- DOOR – Doorway
- DUCT – Duct
- FENCE – Fenceline
- GATE – Gateway
- TASK – Task at critical asset
- PORTAL – Series of two barriers with area between (could be gates, doors, air lock)
- SURFACE – Could be wall, roof, floor
- TUNNEL
- WINDOW

EXAMPLE

In Chapter 7, "System Effectiveness," an example most-vulnerable scenario was used to demonstrate the process. The most-vulnerable scenario was: The adversary enters the facility via the pedestrian gate, crosses the property area and enters the building via the personnel door that is unlocked via working hours. The adversary then enters the control room via force and destroys the control equipment to disrupt the mission of the building. The path elements and physical protection features that are associated with the most-vulnerable scenario include:

- Gate (pedestrian)
- Property Area
- Door (pedestrian)
- Building interior
- Door (Control Room)
- Task (Destroy Control Room equipment with explosives)

	Path Element	Detection	Delay	Selected Path Element
Layer 1	Vehicle Gate	No features	Normally locked Wrought iron gate	Pedestrian Gate
	Fence	No features	5 ft. wrought iron	
	Pedestrian Gate	No features	Always open	
	Area: Transit distance: 100 ft.			Property Area
Layer 2	Vehicle Door	No features	Metal roll up door Locked off hours	Pedestrian Door
	Surface	Personnel during working hours	Reinforced block walls	
	Pedestrian Door	Receptionist during working hours Door alarmed off hours	Tempered glass door Key locked off hours	
	Area: Transit distance: 50 ft.			Building Interior
Layer 3	Surface (control room)	Control room manned 24/7	Framed sheetrock walls	Control Room Door
	Door	Badge reader Door switch alarm Control room personnel 24/7	Hollow-core metal door Electromagnetic strike lock	

Path Element	Detection	Delay	Selected Path Element
Area: Transit distance: 5 ft.			Control Room
Task	Control room manned 24/7	No delay features	Task

Instructions

1. Estimate detection and delay levels for the pedestrian gate:
   a. Detection, Figure C-1
   b. Delay, Figure C-2

2. Estimate traversal time for the Property Area, Figure C-3
3. Estimate detection and delay levels for the Building Pedestrian Door:

   a. Detection, Figure C-4
   b. Delay, Figure C-5

4. Estimate traversal time for the Building Interior Area, Figure C-6.
5. Estimate detection and delay levels for the Control Room Door:

   a. Detection, Figure C-7
   b. Delay, Figure C-8

6. Estimate traversal time for the Control Room Area, Figure C-9.
7. Estimate detection and delay levels for the Task:

   a. Detection, Figure C-10
   b. Delay, Figure C-11

8. Complete Figure C-12:
   a. Identify the first Medium Detection: Pedestrian Door.
   b. Accumulate Delay time after the Pedestrian Door: 28.5 s.
   c. Compare accumulated delay time to response force time.
   d. Accumulated delay time is shorter than the response time, hence, system effectiveness is judged to be *Low*.

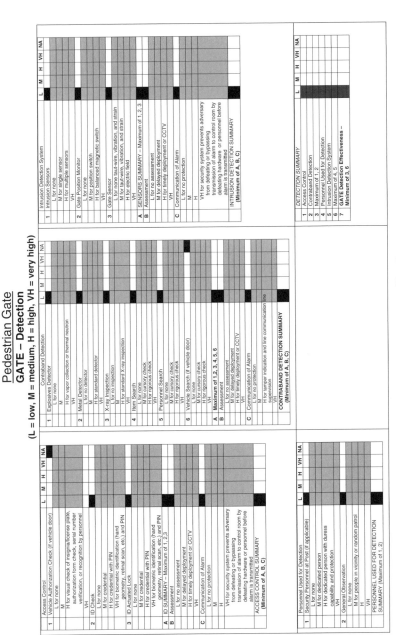

Figure C.1 Estimating Pedestrian Gate Detection Effectiveness.

<u>Pedestrian Gate</u>

GATE – Delay

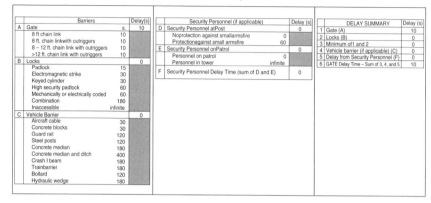

	Barriers		Delay(s)
A	Gate	s.	10
	8 ft chain link	10	
	8 ft. chain link with outriggers	10	
	8 – 12 ft. chain link with outriggers	10	
	>12 ft. chain link with outriggers	10	
B	Locks		0
	Padlock	15	
	Electromagnetic strike	30	
	Keyed cylinder	30	
	High security padlock	60	
	Mechanically or electrically coded	60	
	Combination	180	
	Inaccessible	infinite	
C	Vehicle Barrier		0
	Aircraft cable	30	
	Concrete blocks	30	
	Guard rail	120	
	Steel posts	120	
	Concrete median	180	
	Concrete median and ditch	400	
	Crash I beam	180	
	Trainbarrier	180	
	Bollard	120	
	Hydraulic wedge	180	

	Security Personnel (if applicable)		Delay (s)
D	Security Personnel atPost		0
	Noprotection against smallarmsfire	0	
	Protectionagainst small armsfire	60	
E	Security Personnel onPatrol		0
	Personnel on patrol	0	
	Personnel in tower	infinite	
F	Security Personnel Delay Time (sum of D and E)		0

	DELAY SUMMARY	Delay (s)
1	Gate (A)	10
2	Locks (B)	0
3	Minimum of1 and 2	0
4	Vehicle barrier (if applicable) (C)	0
5	Delay from Security Personnel (F)	0
6	GATE Delay Time – Sum of 3, 4, and 5	10

Figure C.2   Estimating Pedestrian Gate Delay Time.

<u>Property Area</u>

Worksheet: Area Traversal Time

Mode	Distance in Feet	Rate	Delay Time (s) (Distance divided by rate)
Walking		7 ft/s	
Running	100	15 ft/s	7
Crawling		4 ft/s	
Climbing (up or down)		1 ft/s	
Driving (pick up)		54 ft/s	

Figure C.3   Estimating Property Area Traversal Time.

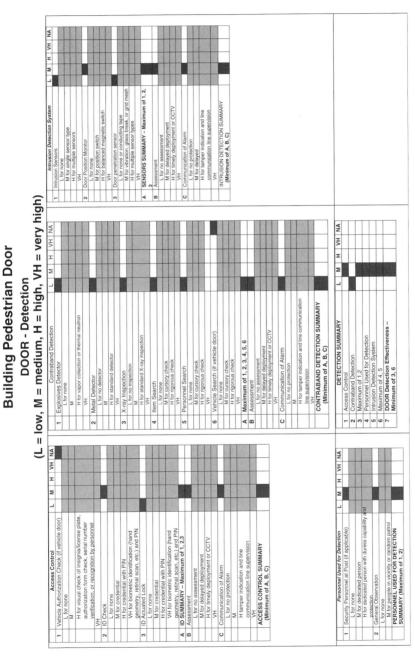

Figure C.4   Estimating Pedestrian Door Detection Effectiveness.

Building Pedestrian Door

DOOR – Delay

	Barriers	Delay (s)			Security Personnel (if applicable)	Delay (s)				DELAY SUMMARY	Delay (s)
A	Door	s. 30		C	Security Personnel at Post	0			1	Door (A)	30
	Wood	10			No protection against small arms fire 0				2	Locks (B)	30
	Hollow core	10			Protection against small arms fire 60				3	Minimum of 1 and 2	30
	Wiremesh	30		D	Security Personnel on Patrol	0			4	Delay from SecurityPersonnel (E)	0
	Tempered glass	X30			Personnel on patrol 0				5	DOOR Delay Time – Sumof 3, and4	30
	Security glass	120			Personnel in tower infinite						
	Steel plate	120		E	Security Personnel Delay Time	0					
B	Locks	30			(sum of C and D)						
	Padlock	15									
	Electromagnetic Strike	30									
	Keyed cylinder	X30									
	Mechanically or electrically coded	60									
	Combination	180									
	Inaccessible	infinite									

Figure C.5   Estimating Pedestrian Door Delay Time.

Building Interior Area

Worksheet: Area Traversal Time

Mode	Distance in Feet	Rate	Delay Time (s) (Distance divided by rate)
Walking		*7 ft/s*	
Running	50	*15 ft/s*	**3.5**
Crawling		*4 ft/s*	
Climbing (up or down)		*1 ft/s*	
Driving (pick up)		*54 ft/s*	

Figure C.6   Estimating Building Interior Traversal Time.

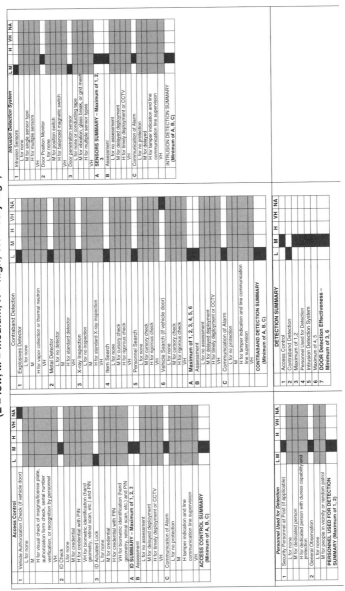

Figure C.7 Estimating Control Room Door Detection Effectiveness.

Control Room Door
DOOR – Delay

A	Barriers	s.	Delay (s)
	Doors:		10
	Wood	10	
	Hollow core	X 10	
	Wire mesh	30	
	Tempered glass	30	
	Security glass	120	
	Steel plate	120	
B	Locks		30
	Padlock	15	
	Electromagnetic strike	X 30	
	Keyed cylinder	30	
	Mechanically or electrically coded	60	
	Combination	180	
	Inaccessible	infinite	

C	Security Personnel (if applicable)		Delay (s)
	Security Personnel at Post		0
	No protection against small arms fire	0	
	Protection against small arms fire	60	
D	Security Personnel on Patrol		0
	Personnel on patrol	0	
	Personnel in tower	infinite	
E	Security Personnel Delay Time (sum of C and D)		0

	DELAY SUMMARY	Delay (s)
1	Door (A)	10
2	Locks (B)	30
3	Minimum of 1 and 2	10
4	Delay from Security Personnel (E)	0
5	DOOR Delay Time – Sum of 3, and 4	10

Figure C.8   Estimating Control Room Door Delay Time.

Control Room Area

Worksheet: Area Traversal Time

Mode	Distance in Feet	Rate	Delay Time (s) (Distance Divided by Rate)
Walking	5	7 ft/s	0
Running		15 ft/s	
Crawling		4 ft/s	
Climbing (up or down)		1 ft/s	
Driving (pick up)		54 ft/s	

Figure C.9   Estimating Control Room Area Traversal Time.

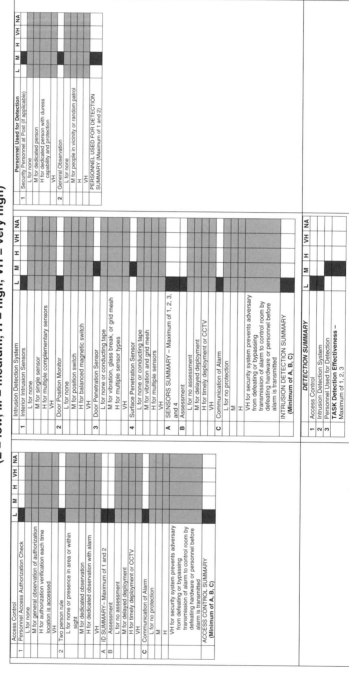

Figure C.10   Estimating Task Detection Effectiveness.

Destroy Control Room Equipment

TASK – Delay

A	Target Enclosure Lock	s.	Delay (s)
	Padlock	40	0
	High security padlock	60	
	Keyed cylinder	45	
	Combination	120	
	Mechanically or electrically coded	120	
	Inaccessible	infinite	

B	Target Enclosure Door		Delay (s)
	Wood	12	0
	9 gauge wire mesh	30	
	Hollow core metal, no lock/hinge prot.	12	
	Hollow core metal	12	
	Tempered glass	5	
	Safety glass	30	

C	Target Enclosure Surface		Delay (s)
	Acrylic plastic	30	0
	Tempered glass	5	
	Polycarbonate plastic	60	
	Laminated glass	60	
	Safety glass	60	
	Wood stud and sheetrock	30	
	Wood stud and plywood	90	
	9 gauge expanded mesh	30	
	16 gauge metal	42	
	¼ in. diameter by 1¼ in. sq. mesh	60	
	3/8 in. by 1¼ in. sq. mesh	300	
	½ in. by 2¼ in. grating	60	
	3/16 in by 2¼ in. grating	60	

D	Target Task Time		15
	Minimal	X15	
	Variable		

E	Security Personnel (if applicable)		Delay (s)
	Security Personnel at Post		0
	No protection against small arms fire	0	
	Protection against small arms fire	60	
	Security Personnel Delay Time (E)		0

	DELAY SUMMARY	Delay (s)
1	Target Enclosure Lock (A)	0
2	Target Enclosure Door (B)	0
3	Target Enclosure Surface (C)	0
4	Minimum of 1 and 2 and 3	0
5	Target Task Time (D)	15
6	Delay from Security Personnel (E)	0
	TASK Delay Time – Sum of 4, 5, and 6	15

Figure C.11   Estimating Task Delay Time.

Estimate Physical Protection System Effectiveness

Path Elements for Most-Vulnerable Scenario	Detection Level	Delay Time (seconds)
Gate (pedestrian)	Low	0
Property Area		7
Door (pedestrian)	Medium	30
Building Interior Area		3.5
Door (pedestrian)	Medium	10
Control Room Area		0
Task at Target	Medium	15
Delay Time After Detection		A: 28.5 seconds
Response Time		B: 300 seconds
Estimated System Effectiveness Level A < B, System Effectiveness = Low A ~ B, System Effectiveness = Medium A > B, System Effectiveness = High		System Effectiveness: Low

Figure C.12   Estimating Physical Protection System Effectiveness.

Appendix D

Insider Threat

INTRODUCTION

The greatest challenge to any security system is protecting against the insider threat. Relative security risk is usually judged to be at the *High* level for the insider because most protection systems are ineffective in preventing an insider from causing the highest consequences *if* they decide to become an adversary. The insider may have authorized access to the building, sensitive information, the information system, and other critical assets. Historically, systematic approaches to address outsider threats have proven to be valuable in developing effective protection systems and for identifying vulnerabilities. An analogous systematic approach must be used to develop an effective protection system for the insider threat and to identify vulnerabilities or gaps in protection. Protection measures for the insider threat may be limited by legal and political issues. High regard must be maintained for adhering to laws that protect personnel privacy and for corporations earning and maintaining the trust of their employees. Security risk will not be estimated here for the insider threat, but instead, the design of an integrated protection system to mitigate the insider threat will be discussed.

Comparable

INTEGRATED PROTECTION SYSTEM FOR THE INSIDER THREAT

What is needed is a security system to mitigate the insider threat that addresses the insider before the employee decides to become an adversary or during the "pre-recruitment" by a malevolent group. Further, the security system must integrate protection functions like personnel security, physical security, cyber-security, and operations security in order to provide protection in depth. Best practices for each protection function cannot just be pieced together with the expectation that the insider threat has been mitigated. A systematic approach is needed to design a performance-based security system to mitigate the insider threat for both intelligence and terrorism concerns. Figure D-1 outlines an approach to develop an integrated protection system to mitigate the insider threat.

The approach builds on five basic steps:

1. Derive undesired events for the insider threat.
2. Analyze the insider threat.

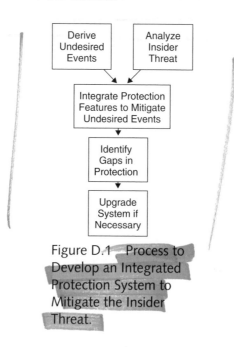

Figure D.1 Process to Develop an Integrated Protection System to Mitigate the Insider Threat.

3. Identify protection features to mitigate undesired events.
4. Identify gaps in protection.
5. Upgrade the protection system, if necessary.

UNDESIRED EVENTS

An initial step in the process is to list all of the possible site-specific undesired events for the insider threat spectrum. These undesired events may include causing loss of mission with the same undesired events that were analyzed for the outsider threat or there may be undesired events particular to the insider threat. Undesired events are those events that you do not want to happen or the undesired events that the protection system should prevent the insider from accomplishing. Examples of undesired events include collaboration with a competitor corporation to compromise proprietary information or recruitment or collusion with an international terrorist group, sabotage of critical assets, or workplace violence. Lists of undesired events will vary depending on the mission of the facility. Undesired events can be ranked or prioritized based on relative consequences.

Each undesired event should be analyzed to determine all of the steps required for the insider to carry out the undesired event, including the recruitment phase or decision to undertake the event, and the actual steps required to successfully complete the event. As was discussed in a previous chapter, a logic tree provides a graphical means to develop the root causes of an undesired event. Development of the tree can be continued to derive all of the means that the insider adversary could use to cause each undesired event. The symbol that appears below the event's name is used to designate either how that event is logically related to other events or how well the causes of the event are known. A reminder is that there are two kinds of logic gates, the AND gate and the OR gate, used in logic diagrams. The shape of the AND gate is a round arch with a flat bottom. For the event described above the AND gate to

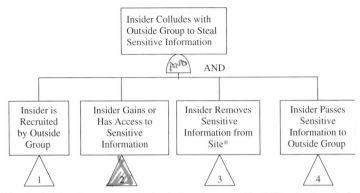

*Note that insider adversary scenarios for theft can be assessed by ASD and path-methods described in Chapter 7, "System Effectiveness."

Figure D.2   Basic Logic Tree for Example Undesired Event.

occur, all of the events that have an input into the AND gate must occur. Thus, if any one of the input events can be prevented, the event described above the AND gate will be prevented. The shape of the OR gate is a pointed arch with a curved bottom. For the event described above the OR gate to occur, any one (or more) of the events that input to the OR gate must occur. All of the input events must be prevented in order to prevent the event described above the OR gate. The transfer operation is represented by an upright triangle. The transfer operation is used to make the graphic display of the logic tree more compact and readable because it allows the tree to "continue" on another page. Figure D-2 provides an example of first-level development for an example undesired event. Note that because all four of the events are required for the top event (the undesired event) to occur, prevention of any one of the four events causes the insider to fail to collude with the outside group to steal information. The process provides a logical, systematic way to use resources effectively. The tree could be developed further in order to derive all of the ways or the scenarios that the insider adversary could use to cause the specific undesired event. Figure D-3 further develops the event "Insider Gains Access to Sensitive Information" represented by the #2 Transfer Symbol.

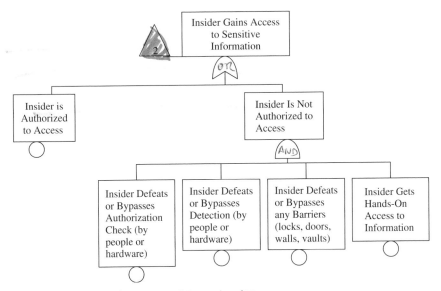

Figure D.3   Development of Branch of Tree.

ANALYZE THE INSIDER THREAT

A description of the insider threat spectrum must be completed in order to design or evaluate an appropriate protection system. It is difficult to know how much protection is adequate without some judgment about the level, access, and sophistication of the threat that the system must protect against. An insider is defined as anyone with knowledge of the operation, sensitive information, and/or the security systems and who has unescorted access to facilities or security interests. A full-range insider threat spectrum would include these categories:

- Passive insider – commits no overt acts
- Provides information only
- Active insider – participates actively
- Nonviolent (unwilling to use force against personnel)
- Violent (active, violent participation – willing to use force against personnel)

The active, violent insider is a very difficult adversary to protect against. More than one insider is possible, but emphasis is placed on addressing the single insider, the most probable insider threat.

The motivations for the insider threat can be the same as those for the outsider threat. Motivation is an important indicator for both the level of malevolence and the likelihood of insider attack. Motivations might include:

- **Ideological** – A fanatical conviction.
- **Financial** – Wants or needs money.
- **Revenge** – Disgruntled employee, contractor, customer.
- **Ego** – "Look what I can do."
- **Psychotic**unstable but capable.
- **Recruitment or coercion**or family or self threatened.

Insider adversaries have advantageous characteristics that distinguish them from other adversaries:

- Operational/system knowledge that can be used to their advantage
- Authorized access to the facility, information system, sensitive information, security systems without raising suspicion of others
  - Can conduct test and rehearsals
  - Can test the system with normal "mistakes"
- Opportunity to choose the best time to commit an act or can extend acts over a long period of time
- Capability to use tools located at work location site
- Recruitment/collusion with others, either insider or outsiders

All employment positions at a facility should be included in the threat analysis. Any employee may pose a potential insider threat, even trusted managers and security personnel. Insider positions might include management, regular employees, service providers,

Table D.1 Example Summary of Insider Capability by Position for a Given Undesired Event

Job Category	Knowledge	Access	Authority
Supervisor	High	High	High
Information system administrator	High	High	Medium
Technologist	High	High	High
Security maintenance	Medium	High	Medium

visitors, inspectors, and past employees. Positions at a typical building might include:

- Managers
- Staff
- Information system administrators
- Security personnel
- Administrative staff
- Contractors
- Custodians
- Maintenance personnel
- Vendors
- Past employees
- Visitors

The product of the Insider Threat Analysis is the identification and characterization of potential insider adversaries. The objective is to identify general personnel job categories in terms of *knowledge, access*, and *authority* related to each of the undesired events/related critical assets. Efforts should be made to ensure that all appropriate personnel assignments are included. The goals are to identify what job categories could provide the greatest advantage for the insider adversary intending to cause the undesired events and to understand the potential capabilities of an insider adversary. Table D.1 provides an example table used to summarize the insider adversary spectrum for a given undesired event. Analysts can use the summary tables for undesired events to complete the insider assessment. A qualitative *High, Medium,* or *Low* judgment should be assessed for the level of: knowledge, access to the critical asset, and authority afforded by each job category to cause the undesired event.

INSIDER KNOWLEDGE

The type of insider knowledge that provides a significant advantage to the insider adversary includes knowledge of: security/control features, work schedules and assignments, locations and characteristics of critical assets, specific details of facility operations, known weaknesses, and gaps in protection.

INSIDER ACCESS

Insider access that can be used to cause an undesired event includes the usual authorized work access, special temporary access to other areas, and the access to other employees as a source of expanded information.

INSIDER AUTHORITY

Insider authority that the insider adversary can exploit is described as management authority over others, personal influence over others, the authority to do assigned tasks, and the ability to get temporary authority to do any task.

PROTECTION FEATURES TO MITIGATE THE INSIDER THREAT

After the insider positions have been summarized in terms of knowledge, access, and authority afforded to the position to cause a given undesired event, the next step is to address the protection system features to mitigate the insider threat. The basic functions of an integrated system to mitigate the insider threat include:

- Minimize the potential for hiring an adversary.
- Deter the on-staff employee from becoming an adversary.

MINIMIZE POTENTIAL FOR HIRING AN ADVERSARY

The natural desire is to not hire anyone with a potential of becoming an insider adversary. Even though there is never a guarantee of this situation, pre-employment screening and the deterrence provided by individuals knowing that they will be screened provides some level of protection. Pre-employment screening could include not only a thorough application process but also some level of background check.

The application process should be very straightforward, and the fact that a background check is a required part of the application process should be very clear. A medical examination and drug test should be required. The application form should be extensive enough to ensure that it asks for all of the information needed to evaluate the applicants. Job opening and application details should be posted far enough in advance to allow time for the background check to be completed before hiring.

Background checks can be as extensive as needed, depending on the level of consequences that could result from the compromise of the facility due to the actions of an insider threat. For some facilities, just an application form and an interview would suffice; others may use some combination of a search of national criminal records, a cursory follow up of the information on the application, and a rigorous follow-up of activity of most recent years. The follow-up might include interviewing references, investigating the candidate's financial affairs, and interviewing previous employers and colleagues. Some level of background check should be repeated on a prescribed schedule to provide active continuous monitoring of personnel.

There are various benefits of conducting background checks. References may reveal information not provided on the application. Criminal records could provide some history of malevolent behavior. Financial history might provide an indication of stability as well as potential susceptibility to extortion. A review of work history could reveal tendencies to anger, reliability, competence, and personal conduct.

DETER EMPLOYEES FROM BECOMING AN ADVERSARY

Several layers of protection should be implemented to deter the existing employees from becoming an insider adversary and to prevent the insider adversary from causing the undesired events. The desired perception is that any malevolent act will be detected and prosecuted. The protection goal for the insider threat is to make it "easy to do the right thing, very difficult to do the wrong thing." According to Turner and Gelles in *Threat Assessment, A Risk Management Approach*, one of the most frequently offered rationalizations by convicted trusted insiders was that "security was lax; tighter security would have been more of a deterrent." The layers of protection to deter the insider adversary might include:

- Security awareness
- Personnel screening for persons in high-risk positions
- Minimization of opportunity for malevolent acts
- Integration of effective security function features
- Proper response to malevolent acts that do occur

SECURITY AWARENESS

Security awareness is a program intended to utilize the non-malevolent majority of employees to detect and deter malevolent conditions. The program is normally part of a routine frequent employee training on operations security. Awareness sensitizes employees to watch for and to report or interfere with potential malevolent actions. Security awareness must address and instruct for the full spectrum of security functions – personnel, physical, cyber-based, information, and so on.

PERSONNEL SCREENING FOR PERSONS IN HIGH-RISK POSITIONS

A higher level of personnel screening may be required for high-risk positions. High-risk positions are those that afford employees access to the most sensitive information or critical assets as a part of their normal job assignment. The additional screening conducted for these persons could include frequent drug screening, physical and mental evaluations, law enforcement checks, credit checks, and supervisor and coworker observation. Screening of persons in high-risk positions must be continuous and timely.

MINIMIZATION OF OPPORTUNITY FOR MALEVOLENT ACTS

Each facility should identify ways to minimize the opportunities for an insider to conduct a malevolent act. A common method is to compartmentalize the facility. Compartmentalization can be achieved by limiting access to sensitive information or actual assets to only those needing it for job duties, further restricting access to the assets that could result in a high-consequence undesired event, enforcing a multi-person presence in critical areas, and allowing for monitoring of activities to detect potential malevolence and to identify who is responsible for the act.

INTEGRATION OF EFFECTIVE SECURITY FUNCTION FEATURES

Protection features from personnel security, physical security, cyber-security, and operations security must be integrated to mitigate the insider threat. Usually, these protection features function independently and are not integrated toward a common objective, but no single one of these functions, acting alone, can answer the insider threat. The protection features must function together to detect, delay, and appropriately respond to the insider threat.

Malevolent actions can be detected by administrative controls, technology controls, and noting suspicious incidents of noncompliance with procedures. Administrative controls might include multi-person presence, key control, and behavior observation. Technology controls might include item or process monitoring, information system use, or communication system use (telephone, e-mail, FAX). A higher incidence of noncompliance with procedures could be "tests" of the protection system or enforcement system. It is extremely difficult to detect the insider who is working with an outside malevolent group. Historically, spies are not usually caught; they are reported by other spies.

Entry/exit control can be used in physical security to enforce compartmentalization and access. Entry control enforces authorization checks with a picture badge inspection, electronic credential, personal identification number (PIN), or some biometric check like fingerprints, eye-retinal pattern, and the like. Entry control schemes might include contraband detection to prevent the introduction of weapons, explosives, or other tools that could be used in a malevolent attack. Exit control is used to detect the unauthorized removal of high-value assets.

Cyber-security measures can be implemented to detect misuse or unauthorized activity with information systems. The monitoring capabilities of information security could report changes in use habits or unusual cyber-activity such as: an increase in accessing critical assets or sensitive information, attempts to gain access to unauthorized computer sites, extensive communication to web sites and/or electronic addresses that are not work related, sending and receiving encoded messages, attempts to access the computer of coworkers, and so on.

The goal of operations security is to prevent sensitive operational information from being inadvertently released. Control is imposed on handling and disposition of hard-copy, electronic media, and any other media containing sensitive or intellectual property. The operations security program can be enforced by monitored procedures and is most effective where security awareness is at a high level. An important component of operations security is extensive and continual employee training in site security matters.

PROPER RESPONSE TO MALEVOLENT ACTS

A precedence of consistent, proper response to malevolent acts may deter the insider considering some malevolent act. The perception among employees should be that malevolent acts will be detected in a timely manner, the malevolent task will be difficult to accomplish, and there will be consequences (prosecution). The response to any malevolent incident should be immediate, the incident should be reported, the perpetrator should be interrogated, and the punishment commensurate with the act.

IDENTIFICATION OF GAPS IN PROTECTION

Once the logic tree for each undesired event has been developed as completely as possible, protection features of the existing protection system can then be associated with the basic events or "bottommost" events of the tree. Logically, if these basic events can be prevented from occurring, the topmost event or the undesired event can be prevented from occurring. Protection features for these basic events can be features from personnel security, physical security, cyber-security, and operations security. The common objective supported by a logic tree is that the goal is to prevent the undesired event. For each basic event, protection features from any or all of the functions should be integrated with each other to prevent the undesired event from occurring. Figure D-4 provides an

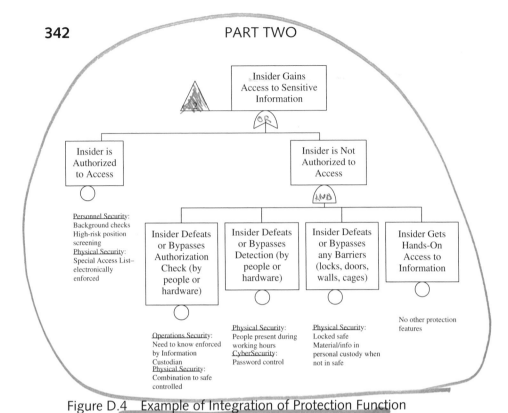

Figure D.4   Example of Integration of Protection Function Features.

example of integrated protection features from various protection functions that might be expected to prevent the undesired event: an insider gaining access to sensitive information.

After protection features have been associated with the events, the next step is to systematically review the features to assess their adequacy in ultimately preventing the undesired event. Gaps in protection are identified by no features or features judged to be inadequate for each event and then for upper events and, finally, for the topmost event (undesired event). The first pass through the logic tree is to identify the basic events for which the protection system provides NO protection, then the analytic team should evaluate the protection system features for the remaining events and judge whether or not the features could perform together to prevent the undesired event. Note that to prevent items with an OR gate, all events pictured below the gate must be prevented since any one of them could cause the event. To prevent items with an AND gate, only one of the events pictured below the gate must be prevented, since all would be required to cause the undesired event. Normally, not one single protection function can adequately protect the event, but through integration and coordination the protection functions can work together to either prevent the undesired event or make it very difficult for the insider to accomplish without being detected.

Protection vulnerabilities or gaps in protection are identified by the lack of protective measures for a given basic event and/or for features that are judged to be inadequate for a particular basic event that logically leads to the undesired event occurring. The identified gaps can occur in one or more of the protection functions of operations security, physical security, cyber-security, or personnel security.

UPGRADE THE PROTECTION SYSTEM

If gaps in protection are identified, the protection system can be upgraded by deriving features to be added for the individual events that would deter the insider or (logically) prevent the undesired event. The process should be continued until all gaps in protection are reduced. The systematic approach provides assurance that the protection functions are integrated to deter the insider and/or prevent the undesired events. Protection features are selected for their function in preventing undesired events. The resultant protection system is based on the integration of protection systems to prevent or mitigate the undesired event.

SUMMARY

The insider threat continues to pose the greatest challenge to protection systems. A systematic approach supports the design of a cost-effective, integrated protection system to mitigate the insider threat. A systematic approach ensures that that the protection functions perform together to mitigate the undesired events and thus make it difficult for the insider to do the wrong thing. It also would begin the detection of the insider threat before the "recruitment by malevolent group" phase. The resultant security system to mitigate the insider threat would integrate all of the protection functions in order to provide a system that is performance-based, rather than compliance-based and to provide protection in depth. In addition, the analysis results would be traceable and repeatable.

REFERENCES

1. Biringer, Betty, *White Paper: Integrated Protection System to Mitigate the Insider Threat,* Sandia National Laboratories, Albuquerque, NM, June 2005.
2. Chapter 22, "Insider Analysis," *The Nineteenth International Training Course for the Physical Protection of Nuclear Facilities and Materials*, Sandia National Laboratories and the International Atomic Energy Agency, April 30 – May 19, 2006, Albuquerque, NM.
3. Turner, James T., PhD and Gelles, Michael G., PsyD, *Threat Assessment A Risk Management Approach*, The Haworth Press, Binghamton, NY, 2003.

EXERCISES

1. List possible security-related undesired events for a building that an outsider adversary and an insider adversary might have in common. What are some undesired events that would apply to the insider adversary only?
2. What are some of the sensitivities and restrictions in protecting against the insider threat?
3. Discuss the three factors for assessing capability associated with insider job categories for causing an undesired event? Are they all of equal importance? Why or why not?
4. What are the two basic functions of an integrated protection system to mitigate the insider threat? How might each be accomplished?
5. Why is it so difficult to protect against a possible insider threat?

Acronyms

ANFO	Ammonium nitrate and fuel oil
ASD	adversary sequence diagram
ASSESS	Analytic System and Software for Evaluating Safeguards and Security
ATF	United States Treasury Bureau of Alcohol, Tobacco and Firearms
BPA	Bonneville Power Administration
CBR	chemical-biological-radiological
CCTV	closed-circuit television
DHS	Department of Homeland Security
EASI	Estimate of Adversary Sequence Interruption
EOC	emergency operations center
FBI	Federal Bureau of Investigation
FERC	Federal Energy Regulatory Commission
HVAC	heating, ventilation, and air conditioning
IFIP	Interagency Forum for Infrastructure Protection
JCATS	Joint Combat and Tactical Simulation
JTTF	Joint Terrorist Task Force
NSTL	National Security Threat List
PPS	physical protection system
SAVI	Systematic Assessment of Vulnerability to Intrusion
SCADA	supervisory control and data acquisition
SNL	Sandia National Laboratories
STAC	Science-based Threat Analysis and Countermeasures

TNT	Tri-nitro toluene
TVA	Tennessee Valley Authority
UPS	uninterrupted power supply
USACE	United States Army Corp of Engineers
USBR	United States Bureau of Reclamation
WAPA	Western Area Power Administration

Glossary

Adversary strategy – Overall plan used to achieve the adversary's objective under advantageous conditions.

Adversary strategy, most-vulnerable – The adversary strategy to which the security system is most vulnerable. The *most-vulnerable strategy* is the one most advantageous for the adversary to pursue in order to achieve the undesired event.

Adversary tactic – Employment of available means to prevent a system feature from accomplishing its purpose. The feature may be part of the security system or a critical asset.

Consequences – Losses to a facility and the public resulting from the defeating of a mission objective. Consequence of loss may be measured in dollars, lives lost, or other measures, but should be consistent to allow for meaningful comparisons. Some consequences may be difficult to quantify, such as political damage or loss of public trust.

Critical assets – Those assets that are essential to meeting the mission objectives. Security systems are intended to ensure that the mission continues to be performed despite malevolent intervention by humans. Identification of the critical assets is necessary before designing, evaluating, or upgrading a security system for their protection.

Defeat method – See *Adversary tactic*.

Delay – A feature that impedes an adversary's progress.

Design threat – The threat against which the security system upgrade will be designed to be effective. Constraints on resources may result in the design threat being less than the maximum credible threat to the facility. Availability of additional resources for security upgrades at a later time might enable a more severe design threat to be adopted. The design threat describes the number of adversaries, their modus operandi, the type of tools and weapons they would use, and the type of events or acts they are willing to commit.

Detection – The sensing, reporting, and assessment of an adversary action.

Domestic terrorist – An individual or group based and operating entirely within the United States and Puerto Rico without foreign direction and whose acts are directed at elements of the U.S. government or population.

Fault Tree – A graphic, logical representation of the relationship among the mission objectives of the facility and the critical assets that support the objectives. A fault tree is built from the adversary's point of view, describing events that cause the facility to fail to meet its objectives by misusing, disabling, or destroying critical assets.

Intelligence – Information and knowledge obtained through observation, investigation, analysis, or understanding. The security system needs intelligence about adversaries. Adversaries need intelligence about the security system.

International terrorist – A person or group of persons who commit an unlawful use of force or violence against two or more nations to intimidate or coerce a government, the civilian population, or any segment thereof, in furtherance of political or social objectives.

Path – Route taken by an adversary from off-site through areas and path elements to reach the target and, optionally, to return off-site. A path is part of a scenario.

Professional expertise/judgment – The knowledge accumulated by trained, experienced personnel employed by the owner/operator of the facility being assessed, especially the knowledge possessed by on-site personnel.

Protection system – Physical security and cyber-security measures used to counter mission threats and consequences.

Response – Interruption and neutralization of the adversary by security police.

Risk – A measure of the uncertainty of achieving a goal or fulfilling a mission.

Risk is quantified by the following equation:

$$P_A \times C \times (1 - P_E) = R$$

where:

$$P_A = \text{Likelihood of attack}$$

$$C = \text{Consequence of the loss from the attack}$$

$$P_E = \text{Security system effectiveness against the attack}$$

$$(1 - P_E) = \text{Security system ineffectiveness}$$

$$R = \text{Risk associated with adversary attack}$$

Sabotage – Whoever, with intent that his or her act shall, or with reason to believe that it may, injure, interfere with, interrupt, supplant, nullify, impair, or obstruct the owner's or operator's management, operation, or control of any agricultural, stock-raising, lumbering, mining, quarrying, fishing, manufacturing, transportation, mercantile, or building enterprise, or any other public or private business or commercial enterprise, wherein any person is employed for wage, shall willfully damage or destroy, or attempt or threaten to damage or destroy, any property whatsoever, or shall unlawfully take or retain, or attempt or threaten unlawfully to take or retain, possession or control of any property, instrumentality, machine, mechanism, or appliance used in such business or enterprise, shall be guilty of criminal sabotage.

Scenario – Outline of events along a specific path by which the adversary plans to achieve his objective.

Scenario, most-vulnerable – The (adversary) scenario that takes the greatest advantage of the vulnerabilities of the security system.

Target – A point, object, or goal at which something else is directed. A critical asset to be affected (i.e., misused, disabled, or destroyed) by an action of an adversary.

Terrorism – The unlawful use of force or violence against persons of property to intimidate or coerce a government, the civilian population, or any segment thereof, in furtherance of political or social objectives.

Terrorist – An adversary who uses violence, terror, and intimidation to achieve a result.

Terrorist group – A collection of adversaries, commonly working in small, well-organized groups or cells that repeatedly commits acts of violence or threatens violence in pursuit of its political, religious, or ideological objectives.

Threat – Anything that can disrupt the mission of the facility. A facility may face several malevolent threats to its mission. The many varieties of adversaries fall into three classes: insiders, outsiders, and outsiders working in collusion with insiders.

Threat assessment – A systematic evaluation of the threat that identifies and describes the adversaries a facility may face in the future together with an estimate of the likelihood that an attack will occur.

Threat, site-specific – The spectrum of threats to the site being assessed; a subset of the generic threat. The site-specific threat may be common to several sites within a geographical area.

Vulnerability – A weakness or gap in the protection system.

Index